T0074588

Hark, Hark! Hear the Story of a Science Educator

Global Science Education

Series Editor:
Professor Ali Eftekhari

Learning about the scientific education systems in the global context is of utmost importance now for two reasons. First, the academic community is now international. It is no longer limited to top universities as the mobility of staff and students is very common even in remote places. Second, education systems need to continually evolve to cope with the market demand. Contrary to the past when pioneering countries were the most innovative ones, now emerging economies are more eager to push the boundaries of innovative education. Here, an overall picture of the whole field is provided. Moreover, the entire collection is indeed an encyclopaedia of science education and can be used as a resource for global education.

SERIES LIST

Hark, Hark! Hear the Story of a Science Educator

Jazlin Ebenezer
Wayne State University

CRC Press
Taylor & Francis Group
Boca Raton London New York

CRC Press is an imprint of the
Taylor & Francis Group, an **informa** business

First edition published 2020
by CRC Press
6000 Broken Sound Parkway NW, Suite 300, Boca Raton, FL 33487-2742

and by CRC Press
2 Park Square, Milton Park, Abingdon, Oxon, OX14 4RN

© 2020 Taylor & Francis Group, LLC

CRC Press is an imprint of Taylor & Francis Group, LLC

Library of Congress Cataloging-in-Publication Data
Names: Ebenezer, Jazlin V., author.
Title: Hark, hark! Hear the story of a science educator / Jazlin Ebenezer.
Description: First Edition. | Boca Raton ; London : C&H/CRC Press, 2020. |
Series: Global science education | Includes bibliographical references
and index.
Identifiers: LCCN 2020001069 | ISBN 9780367224172 (Hardback) |
ISBN 9780429274763 (eBook)
Subjects: LCSH: Science teachers—United States—Biography. |
Science—Study and teaching—United States.
Classification: LCC Q143.E35 A3 2020 | DDC 509.2 [B]—dc23
LC record available at https://lccn.loc.gov/2020001069

ISBN: 978-0-367-22417-2 (hbk)
ISBN: 978-0-429-27476-3 (ebk)

Typeset in Times
by codeMantra

Hark, Hark! Hear the Story of a Science Educator is dedicated to all individuals who thrill my mind, heart, soul, and spirit.

- Dr. D.L. Ebenezer, my husband, who encouraged me to become an educator
- My father, S.T. Jabez, and mother, Jane Jabez, who passed on their genes to me, their oldest daughter. They have pointed me to the power of purpose, passion, patience, perseverance, and perpetuity to excel. My creed is "Whatsoever your hands find to do, do it with all your mind, heart, and soul"
- My mentors who nurtured me and pointed me to the path to success
- Science educators and all the readers who find my academic memoir a source of inspiration that illuminates their scholarly path
- My four-year-old granddaughter, Celina, who inspires me with her statements

Contents

Foreword

Hark, Hark! Hear the Story of a Science Educator is probably the first autobiographical publication on science teaching and learning devoted to the dissemination of the tenements of the latest research on the teaching and learning of science to the understanding of the public. This labor of love was undertaken by a damsel who holds a votive candle for the popularization, propagation, and promotion of science education among masses. Unbelievable is her global pan-vision, vast and varied experience, rare and unique expertise, enviable patience, adamant persistence, and incredible perseverance and dedication and devotion to science education. The exciting episodes of her academic life depict a gradual but rare evolution of a science educator from an unlikely background. It was an unending series of efforts, but she successfully surmounted them all with patience and perseverance to reach this hallowed spot. The consequence is the numerous awards, rewards, citations, and designations but only cropping up on the sidelines, and being in the enviable place where she is right now as the *Queen of Science Education* as someone described her. This extraordinary book, an academic and literary marvel with its exciting and captivating chapters, but carefully written, gradually unfolds the wonders of science education. May it inspire its readers, the public, as well as the university, college, and school teachers and students who tread the realms of post-modern science education.

S Rajendra Prasad, M.Sc.
Retired Professor of Chemistry and Education and
Registrar of Spicer Adventist University, Pune, India

I have never read a book for educators filled with such wisdom, care, conceptual learning, and God's grace weaved throughout. *Hark, Hark! Hear the Story of a Science Educator* is a treasure, and I am grateful for its creation and gift to future educators, researchers, and graduate students. The author is a prolific writer, researcher, storyteller, and dynamic teacher. Her story of a female science educator and researcher traversing academia for K-12 and university institutions is compelling. Her documented story describes her quest to fulfill God's will with excellence, determination, and grace. Her story is the foundation for her commitment to teaching, researching, and writing.

The author reflects her lifelong passion and perseverance to excel. With grace, she gives her mind, heart, and soul to working with her students whom God has placed under her wings.

Lynda Wood, PhD
Retired Superintendent
Southfield School District, Southfield, MI, USA

Preface

Hark, Hark! Hear the Story of a Science Educator is meant to empower the public understanding of science teaching and learning. The book is the outgrowth of my desire to someday write a book that even the airports will market in their bookstores to worldwide travelers. Situated in a personal narrative, this book will appeal to a broader readership. The seed to this book was sown by Dr. Beata Biernacka, University of Winnipeg, Winnipeg, Manitoba, Canada. As my PhD student in science education at the University of Manitoba, she said, "You need to write your academic experience that shows how a woman in science education navigates the system."

Chapter 1 begins with my academic journey at the University of British Columbia, Vancouver, Canada, where I received my doctorate in science education. It reveals how I established my scholarly roots in research, teaching, and service. It begins from me not understanding the academic papers I read and ends with the rewarding experience of presenting my first paper at the 1989 Biennial Chem. Ed. Conference at Queens University, Canada. I also highlight the struggles in my academic pursuit. I provide two knowledge construction tools advocated by Joseph Novak to those who wish to understand science content. I have been teaching these knowledge construction tools for nearly 35 years to undergraduate and graduate students. These tools are useful, even in planning and evaluating research.

Chapter 2 continues my story, still at the University of British Columbia. I narrate how a visiting professor calls me out boldly to engage in academic writing by making an innocent statement, "You are too lazy." I trace the passion for writing to Lady Doak College, Madurai, India, where I contributed articles to the college journal without an invitation. I give credit to Professor Uri Zoller, Haifa University, Haifa, Israel, for resurfacing my talent. I point out how Professor Gaalen Erickson, University of British Columbia, Vancouver, British Columbia, my advisor, endorsed my writing. He stated to a master's student, "Jazlin is adept at writing." I also discuss how Linda Montgomery, editor of my teacher education textbooks, said, "Your writing is fluid." I highlight how another professor claimed that my writing was too soft for teachers. In Chapter 2, I also talk about my research in the classroom. What is learned in the qualitative research course plays out vividly. This chapter also helps the readers infer what is most important in academia. Publishing ought to begin in doctoral course work. A section admonishes that publications from dissertation should waste no time. It points out that the inclusion of advisors as co-authors is to appreciate their involvement in building and shaping my intellectual capacity.

Chapter 3 describes the launch of an academic position. My translation of theory into practice in the elementary preservice science education methods course opens my eyes to several issues associated with teaching. The impressions some preservice teachers created in me are included. The origin of my first two books is part of this chapter. I also narrate the interactions with national and international scholars, which include South Africa and Turkey. Then I move on to my sabbatical in the East and the West. I introduce argumentation for science teaching and learning. I narrate the luxurious World Bank Mission to Indonesia to evaluate the seven-year, $56M project. The call to go on a mission stemmed from an unexpected address on science teaching and learning at the World Bank, Washington, DC.

Chapter 4 uses "The Science Banqueting Table" as a metaphor to discuss the variation theory of learning for relational conceptual change inquiry. Stemming from this theory is the Common Knowledge Construction Model (CKCM) of Teaching and Learning. It is an empathetic model that aspires to reach all students. Other than Canada, where the CKCM originated and the US where currently practiced, the CKCM has an enormous influence in Turkey. Beginning from 2013 until the change to traditional STEM textbooks in 2018, the CKCM underpinned the Turkish primary (grades 3–8) science textbooks funded by the Ministry of National Education. The Turkish scholars use the CKCM for their master's theses and dissertations on teaching and learning science. Besides, there are several Turkish studies published in journals on the effectiveness of the CKCM on the conceptual and affect. The example of my research on the model, its underpinning theory, and the teaching strategy included in this chapter are borrowed from the *Journal of Research in Science Teaching*, a premier journal in science education. I narrate the origin of this article in my interactions with Professor Shiela Chacko, Spicer Adventist University, Pune, India. This conversation was en route to Tamil Nadu by train in 2004 while moving ahead of the tsunami in the Indian Ocean that affected several countries in Southeast Asia. While my family was trying to locate me, I did not panic. To me, it seemed a movie show on India's national TV.

Chapter 5 vividly narrates the events at Wayne State University. I portray that research, teaching, and service are ONE. I describe that Lake Erie Watershed probe with innovative technologies is the prophesied research, which the cover page of my book for secondary teachers depicted. I focus on my "You Reach" trips in China, Thailand, and India. These invited trips give a glimpse into how an academic can influence global scholars with one compelling message using the profound, yet simple, salt-water activity.

Chapter 6 tracks my path from HyperCard to Artificial Intelligence for Relational Conceptual Change Learning. Although my dissertation was not on the use of technology in science teaching and learning, the results of my research during 1987–1990 inspired me to develop computer animations. I also use the virtual platform for discourse. Finally, this chapter outlines the TESI model I created from my research and its influence in Turkey. I am now engaged in developing and testing the Students' Ideas App (SIA) in university and school classrooms. This engagement is with the help of a computer science educator who I came across and did not hesitate to work with, Dr. Pitchumani Angaykanni Sekaran. The funds that I received through my distinguished professor award serve the development of the SIA.

Chapter 7 examines engineering design for learning science content using societal issues. With the help of one of my bioengineering colleagues at Wayne State University and an environmental engineer, I provide examples of societal problems for engineering solutions.

Chapter 8 uses the Pythagorean Theorem as a metaphor for the architecture of curriculum design. The frameworks are borrowed from William Doll, a world-renowned curriculum theorist with whom I have had many conversations.

Author

Jazlin Ebenezer, EdD, Charles H. Gershenson Distinguished Faculty Fellow, is a Professor of Science Education and Curriculum Studies at Wayne State University. Professor Ebenezer has been researching students' conceptions of scientific phenomena and the integration of innovative technologies into relational conceptual change inquiry. Professor Ebenezer is currently interested in machine learning. A computer science educator is collaborating with Dr. Ebenezer to design and develop the Science Ideas App that will categorize students' ideas and integrate them into the science curriculum. She has published numerous scholarly articles and some books.

Her awards include Charles H. Gershenson Distinguished Faculty Fellow in 2019, WSU, Fulbright Specialist Award by the U.S. Department of State in 2017, and 2000 Rh Award Research and Development, Research Merit Award, and Outstanding Students Honoring Outstanding Teachers Award by the University of Manitoba.

Professor Ebenezer was the founder of the Shuswap Seventh-day Adventist School, Salmon Arm, British Columbia, Canada, in 1982. This school is operated by the Seventh-day Adventist Church, British Columbia, Canada. Together with her family, she has educated school and university students at various levels in Bangladesh, India, and Sri Lanka. Professor Ebenezer has established the Dr. Jazlin Ebenezer Research Fund, a Wider Church Ministries endowment fund, in honor of the education she received at Lady Doak College, Madurai, India. The Fund provides a grant to small teams of multidisciplinary faculty members at Lady Doak College to conduct research on how women learn science, as well as to disseminate findings at national and international conferences. She completed her EdD in Science Education at the prestigious University of British Columbia, Canada. To contact her via e-mail: aj9570@ wayne.edu.

Treading the Academic Waters

1

1.1 MY ACADEMIC JOURNEY BEGINS

My adventure as a science educator took roots when I began my doctoral studies in science education in 1987 at the University of British Columbia (UBC), Vancouver, British Columbia, Canada. On the first day of class on a hot September day, I waited, along with five others, for the arrival of a well-dressed professor. But I was quite surprised when in walked the professor with an old loosely knit, brown half-sleeve T-shirt as I did not expect that kind of attire from a science education professor!

We introduced ourselves to the class. The professor said that he would give us papers to read each week and did not say much in class that day. We spent almost all of our time together in silence, except for a few words of exchange. I was anxiously waiting for the first three-hour class to be over.

I started reading the paper that night but did not understand the contents of the article. Being a chemistry undergraduate and taking physical chemistry concentration at masters' level in education, I did not understand anything that I read. I remembered stopping at every university along the west coast of the United States to inquire about a PhD program when I was on a holiday trip with my family. The graduate advisors/professors at each university said that they would admit me when they looked at my academic records. In one university, a professor suggested that I might prefer a doctoral program in education, focusing on general science while adding that a PhD in science education would be challenging. I told myself that if that were the case, I would follow the challenging route. Finally, I am in my first class at UBC experiencing a conceptual struggle with the readings.

Week after week, we would come to our three-hour class after reading the long assigned academic papers. I was the quietest in class because I did not have the disciplinary language or competencies to articulate my thoughts. Nor did others engage in an active conversation. Our professor attempted to join us in a discussion once in a while. Only the First Nation student among us had something useful to say. I often wondered from where she was getting all this wisdom. Later, I realized that the First Nation has a storytelling culture. Of course, I was waiting for the professor to say something more. There was much time spent in silence, but that did not matter to the professor. The discussion revolved around the paper. At the end of the class, I dared not express my feelings to my peers though I thought to myself; I have not learned much. This course was year-long starting in fall and ending in the spring. It was pure drudgery! Reflecting on the professor's orientation to teaching, I understand that he was listening to students' voices keenly. He was allowing the students to take ownership of the class discussion and to express their thoughts without constraints. Waiting patiently without premature commentary is a virtue in teaching!

I wanted Gaalen Erickson, the professor mentioned above, an emeritus retired professor now, to be my doctoral advisor (Figure 1.1). There were two reasons for this choice. His research was on learning. Saroja, the secretary of math and science department said he is internationally known, so be with him.

FIGURE 1.1   D.L. Ebenezer (left), Gaalen Erickson, doctoral advisor (center), and Jazlin (right), 1991 convocation.

Although I looked naïve to the beholder, I was highly political and wanted the best because I have to ride on my advisor's glory at the time of looking for a position. This reason played out later and carried much weight because some said she is Gaalen's student. Later on at an International conference in Canada, listening to my talk, a UBC alumnus said history is repeating.

1.2 TWO KNOWLEDGE CONSTRUCTION TOOLS

In the first semester of this science education course, our professor assigned us to write two papers. One was a book review and the other a literature review on a particular topic. I did a book review on *Learning How to Learn* by Joseph Novak and Bob Gowin (1984), the inventors of the two knowledge construction tools, Concept mapping and Vee diagramming, respectively. I used these tools in a unit on solution chemistry and, subsequently, I submitted the class assignment to the *Journal of Chemical Education* (see Figures 1.2 and 1.3).

These visual knowledge construction tools show a relationship theory and practice that is foundation to the relational conceptual change that is discussed in a forthcoming chapter in this book. I introduce these two tools to all my students for learning science topics, inquiries in science, and educational research. Some Ph.D. students have used Vee Diagramming as an organizational tool to represent their research, connecting the conceptual and methodological.

1.3 CLASSROOM DISCOURSE

In the second half of the year-long course, as part of our assignment, we observed a teacher, took field notes, and wrote a paper. I watched a chemistry teacher at the university prep school behind my apartment for two days. The chemistry teacher's instruction was on the types of chemical reactions. My professor asked me to use the book *Common Knowledge: The Development of Understanding in the Classroom* authored by Edwards and Mercer (1987), professors at the University of Cambridge, London, England, to interpret my observation. I used the book as a lens to frame the class discourse on the type of chemical reactions, which I audio-recorded on two successive days

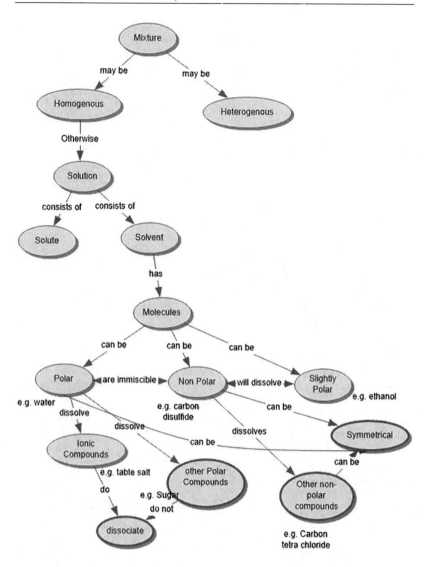

FIGURE 1.2 A concept map representing the key concepts and propositions taken from a unit on solution chemistry. A concept map to illustrate how progressive differentiation and integrative reconciliation might take place in the cognitive structure of a student when he/she engages in further inquiry of the general concept "solvent" chemistry. (Ebenezer, 1992, 465.)

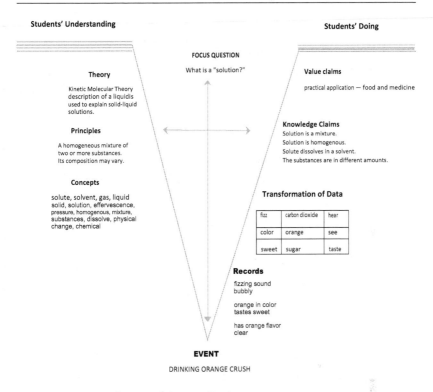

FIGURE 1.3   Vee Diagram of Orange Crush.

and transcribed verbatim. The chemistry teacher's instruction pattern heavily involved fill-in-the-blanks and cued elicitations.

I creatively wrote a paper on the classroom discourse because I also read related articles and books that Edwards and Mercer had cited. My professor liked the paper and wanted a copy for his file. I was pleased and excited. I then asked him what I should do to publish this paper. He said that my theoretical base should be firm, and I need more data collected over a long period to support my interpretations. My eagerness to write articles like the ones I read were apparent. I relished composing the two papers. Reading the two books mentioned above was straightforward, unlike the scholarly journal articles. The year slipped by.

1.4 EXPOSURE TO CURRICULUM STUDIES AND RESEARCH METHODS

Along with the science education course, I also took a year-long course in curriculum studies with master's students. We focused on many curriculum theorists and their frameworks. We also read papers on curricular issues at the provincial level though it was not easy for me to fathom any of these.

I also took two semesters of research methods, one on the fundamentals of quantitative and the other on ethnography. I enjoyed the ethnography course. Our course text was *School Experience* by Woods and Hammersley (1980). As part of the course, I observed a scenario and applied the principles of ethnography. In the second year, I took a year-long course on the sociocultural theories of science education. We discussed many papers. It was a struggle to immerse myself in educational research papers, and I felt strongly about my inability to carry on a meaningful conversation or contribute to the class.

1.5 OVERCOMING CHALLENGES TO READING ACADEMIC PAPERS

The challenge of understanding the reading material was mainly because of having come from a traditional science background. No education issues and jargon were involved in lectures and labs at Bishops College, the first private girls' high school in Sri Lanka, or BSc Special Chemistry at Lady Doak College, a prestigious private women's college in Tamil Nadu, India. The bachelor's and master's in education with physical science as the concentration at the University of Washington required reading some books on the educational foundation. I did not enjoy reading and slept through much of the course readings and yet got a letter grade A in most courses. Looking back, I wonder what the difficulty was about any of the course readings. When my students now say that science education and curriculum articles are challenging to read, I can understand why. They say that they do not have sufficient background to understand the ideals of curriculum studies. Some of my students translate to those who are second language learners. Some give the excuse they are English as a second language learners and cannot write well. I tell them that native-borns also find it hard to write academic papers. An internationally renowned science educator, while having dinner stated that our doctoral students are babies in research and we need to nurture them. In contrast, a dean told at a meeting that if students do not know how to write academic papers within the first two years, they should be encouraged to drop out. A professor from a prestigious university has also told me that working with foreign students is challenging because of the language facility. Where do we stand?

I cultivated the art of writing papers by reading profusely and writing articles. One time, a master's student advised by the same professor I referred to earlier was in his office. He told the master's student that I am "adept" at writing and to get help from me to write her thesis. His statement took me by surprise. I was struggling to make sense of the reading materials and write scholarly papers. It was, indeed, a productive struggle. However, I was passionate about writing and wanted to learn the art of writing. I took the time to read each article, observe how the authors organized their papers, and how they phrase, although I did not clearly understand how to write an academic essay. Experiencing conceptual challenges to read and write scholarly papers have helped me to reach out to infants in academic research.

1.6 ATTENDING THE FIRST SCHOLARLY CONFERENCE

There was an advertisement posted about 1989 Chem. Ed. Conference at Queens University in Canada in the Department of Mathematics and Science Education Department, Faculty of Education at UBC. I responded to this call by submitting a proposal based on my pilot data that I had collected in a grade 11 chemistry classroom. The scientific committee of the Chem. Ed. Conference accepted my proposal as a paper presentation. I informed my professor about the acceptance of the proposal. He asked me whether I was going to submit a proposal. I said, "No, it has been submitted and accepted." And then at another meeting with him, I requested travel funds to attend the conference. Shortly after, I received a formal letter from the head of the Mathematics and Science Education Department offering me a travel grant of $200, stating this would be a one-time award.

At the international conference, the researchers said that I should present my paper at the Chemistry Education Research Conference. The chemistry educators saw potential in me as a researcher and were enthusiastic about my research on students' conceptions. At the meeting, I met a nun who was a chemistry educator. I asked her, "How does one publish papers?" She said, "I need to submit the paper, wait patiently, and pray."

The following week in my science education research course, the professor asked about our experience that week. When it was my turn, I told the class about the advice the nun had given me. My peers and the professor had a good chuckle. In my Curriculum Studies class too, I asked, "How can I learn to write?" The professors teaching the curriculum course stated, "No one prepares you to write, but you merely write. Over time, you will learn to write!"

You Are Too Lazy!

2

2.1 MENTORING

During my first year of doctoral studies in science education, I also took a year-long course that focused on science, technology, environment, and society. Uri Zoller, a Jewish professor from Haifa University, Haifa, Israel, who had come on a year-long sabbatical, taught the course. In this class, I was the only student pursuing a doctorate in science education. Except for one visiting scholar from a Southeast Asian country, the other five in the class were full-time science teachers from Vancouver and suburbs. They were doing their masters' program part-time. We wrote two papers on science, technology, environment, and society, one a theoretical, the other on classroom application.

Apart from assignments, the professor wanted us to collect data for his research paper that he had already written. While the others were able to collect data in their school districts, I found a school to collect data. He promised that all the seven students would be co-authors, and the names would be in alphabetical order. The professor plugged the data that the students had gathered. Shortly, he published our paper on goal attainment in science–technology–society education and reality in *Science Education*. It was a case study of British Columbia.

When Uri invited me to conduct research with him, I was pleasantly surprised. Perhaps, he asked me because I was the only full-time doctoral student in his class. I hesitated because I was a novice, beginning to learn about research, and did not think I could help him. It was then that the professor said, "You are too lazy." I did not get offended by or react to his comments that accompanied with a smile. It seemed that he alerted me to something significant. My South African piano teacher who prepared me for the Canadian Royal Conservatory grade eight examinations reminded me often that I need to be "lazy" in playing the piano. In both instances, the persons used the word "lazy" metaphorically in a good sense, the first to wake me up through an academic call, the second to be light-spirited, bouncy, and playful, not weighed down with burdens.

Richmond School district wanted help from the University of British Columbia (UBC) to interview their grade nine students on their attitudes toward science learning. They had already conducted a survey of their attitudes toward science learning. My professor asked me whether I would take on this task. I interviewed 90 grade nine students and transcribed the qualitative data verbatim and wrote a well-documented report and submitted it to the school district. I asked the district authorities whether I can use the data to write articles. The school superintendent was willing to give me both the survey and the interview data. So, who did I seek help with writing? You guessed, right. It was professor Zoller who called me lazy. So I spent that summer learning from him how to write articles. One was about the no change in high school students' attitudes toward science learning even though the British Columbia Department of Education had instituted science–technology–society curricula in its schools. The other was about grade ten students' perceptions of and attitudes toward science learning. Both articles were published in scholarly journals and I was the lead author. The professor, who called me lazy, spent all summer to help me write these articles.

2.2 SCIENCE EDUCATORS' RESPONSES

I mentioned to my advisor that I had published two papers, and he said, "Are you cranking out papers?" Seeing the brochure that announced my doctoral orals with my biography, my peers were in awe. When I requested a professor to write a recommendation letter for academic positions, he stated that my CV looks "padded." None of these comments penetrated my head; perhaps my cerebrum did not register emotions. Probably, my professors thought that it was too early in my program or too premature to publish.

Some years later, Jonathan Osborne, a science education researcher, and educator currently at Stanford University, referred to my article on attitudes. He argued in his 2003 review of the literature on attitudes toward science that attitudes research in science education should use a combination of methods, both quantitative and qualitative, for establishing validity.

During my second year at UBC, I took a course that focused on the study of emotions. I learned how to develop Likert-type survey items and conduct factor analysis using a program. In this course, too, the professor asked for my paper for his collection. Perhaps, I was good at translating theory into practice. Richard Duschl, a professor of science education, currently at the University of Philadelphia, mentioned to me a few years ago that my research is at the interface of theory and practice.

What I did at the UBC as part of my course work developed my character and prepared me for academic life in several ways. One was to write for publication while still in the doctoral program. In other words, to consider each course as a platform to do research and disseminate.

Recall in Chapter 1 how I published my first paper in the *Journal of Chemical Education* by doing a book review in the first course I took. Next, to be willing to participate with a research expert or a group to conduct investigations and write papers. Although some narrow-minded students are of the view that they are helping professor to publish, I view it as an opportunity to develop the craft of writing.

2.3 RESEARCH ROUTE

My advisor introduced us to the literature on constructivism, which focused on learning. I was very keen on student learning in chemistry because I had earned a degree in BSc Chemistry from Madurai University, Tamil Nadu, India, in 1970. Based on the constructivist literature I read, I did a pilot study in a grade 11 chemistry class at Richmond High School, Richmond, British Columbia, Canada to refine my interview questioning techniques. This classroom chemistry teacher was the one my advisor asked me to help write her master's thesis. My advisor put us together so that we help each other in our work. I collected pilot data in her classroom with a few students.

Subsequently, when the time came to gather data for my dissertation, I interviewed all 13 students individually, 45 minutes each at their convenient time. When I gave the data and encouraged her to use it to teach solution chemistry, the teacher invited me to teach her class. But she was not happy when I moved tables and chairs and set up demonstrations. I could not understand why as students were thrilled.

During the same period of instruction in her class, the teacher was continuing with her master's course work. One day when my advisor had to go somewhere, another professor took his class. At that time, part of the class discussion was on research, data collection, and who owns the data. After this class, the teacher preferred not to have me in her class. Our advisor came to my office and said, "We need to get back." It appeared that he had already talked to the teacher about the problem. She was willing to have me back in her class.

She had also conveyed to him that I was a good chemistry teacher. Despite these kind comments, I was reluctant to go back to her classroom.

I transcribed the individual interviews of 13 students based on their conceptions of three chemical systems: salt in water, saturated sugar solution and crystal formation, and immiscibility of oil in water. I asked my mother to transcribe the tapes on classroom teaching. Each hour of audio-recording took my mother and me more than four hours to transcribe. And transcribing the messy classroom discourse data was not easy. The lessons I learned in my ethnography course played out vividly in the chemistry classroom. Access to school, class, teacher, and students; who owns the data; gatekeeping; developing a mutually trusting relationship are all issues I faced. However, I am still a classroom researcher treading softly in the classroom. Moreover, I work with teachers and students with much intellectual empathy. I advise my doctoral students to do similar research. Two science education researchers have told me that my writing based on data from classroom research is soft. Through writing, teachers should be exposed to practice that reflects current research on teaching and learning.

My mother and I took a month to transcribe the interview data. I wrote the methodology chapter first and followed it up with the results chapter, which took two weeks. Then I completed each of the other chapters, one per week. I submitted the dissertation to my advisor and waited patiently for many weeks. He told in our weekly Tuesday group meeting in the "hut", a little old building adjacent to the Faculty of Education building that I am almost through with writing. He finally said, one day, when I met with him in the office, he does not have time to read this year and for me to work at UBC for one year. I was pleased with this offer.

2.4 DRAMATIC IRONY

I had already applied to a one-year temporary position at Florida State University (FSU), and the chair of the search committee had phoned my advisor looking for me that summer. During this time, I was on a trip to Florida with my husband and son. This trip was not unusual. Each summer, we go on trips in our "baby" blue Honda Hatchback. That is how we have traveled to every state in the United States and each province in Canada.

While driving, seeing the sign to FSU, my husband asked me whether I wanted him to drive me to talk to a professor about the job posting. I said, "No one responded, and I do not feel like visiting anyone." It was during the same time that the search committee chair was calling UBC to locate me. I had not left any information about my whereabouts with anyone. My advisor, wishing that the position should go to a UBC student, was quick to suggest another student, who was ahead of me by two years. FSU had called her for the interview. This student told me the exciting news. I was quietly listening and thinking to myself whether it is the same position that I had applied for. Many thoughts crossed my mind.

Later that week, my advisor called me to his office and related the episode. My peer also thanked me for paving the way for her to commence academic work. My advisor told me that it is better that I stay at UBC while he completes reading my dissertation and then look for a position. While waiting at UBC in my fourth year, I took the opportunity to learn Hypercard so that I can create science activities with animations. This learning later became very useful for research and teaching. I learned to believe that everything happens for a reason. I also know that when one door closes, seven others will open. So thoughts of cheer filled my troubled soul.

My moments of gloom soon vanished. I was taking lessons with a retired nurse, learning to craft realistic sugar flowers, wire them, and make them into bouquets. Observing a little basket of flowers, a science educator commented, "You must have been meditating as you crafted these flowers." Sugar crafting also requires to be "lazy." Since those six classes on sugar crafting, I have decorated fascinating wedding cakes for relatives and friends free of charge. The wedding cake I decorated with more than 1000 flowers was the centerpiece at my son's wedding at Fairmont Hotel, Vancouver, BC. A guest at the wedding said that he would be willing to sell the sugar crafted flowers to the celebrities at Hollywood because only a few can afford realistic flowers. The most recent one I decorated was for Dr. Jasmine Jabez (Joseph), my niece's wedding. The reception planner equated the flowers to a New Yorker who creates such flowers.

2.5 DELAYED RECOGNITION

I was a graduate research assistant and graduate teaching assistant at the Faculty of Education. As a full-time graduate student, I had the opportunity to be a lab assistant, teaching elementary and junior science methods courses and supervising students in the practicum. During my first year, I was a lab assistant. In the second year, I taught an elementary methods course. However, there was reluctance having me teach junior science methods and advise preservice teachers during the practicum. I was expected to observe someone supervising the preservice teachers during the internship for several days, which was not required for others. The work supervisor admitted that he was very hard on me throughout the program, and yet I overcame those difficulties and successfully completed the program.

In the fourth year, as planned, I defended my dissertation successfully on the second floor of the administration building at the UBC in the presence of nearly 30 graduate students. In my trial run of the thesis with the science education faculty and peers in the evening before the big event, several classmates asked me where I learned the art of public speaking. I said, "Seventh-day Adventist Church." One of the professors had also told another friend of mine, we did not recognize her talent for speaking. One question during this session was what is the difference between constructivism and phenomenography. I drew two stick figures on the board to distinguish between the two theories of learning. Constructivism was represented by a circle is in the head to represent the schemas. Phenomenography was represented by an arrow between the subject and object to indicate it is the relational aspect that is significant. Chapter 4 describes phenomenography in more detail.

The building in which I defended my dissertation was the same where UBC authorities told me when I first inquired in 1972 about admission to prove myself in a community college. Now I am taking a detour in my storytelling. I crossed the Canada–United States border from Vancouver in 1976 to explore the possibility of doing my masters at Western Washington University, Bellingham, WA, USA. The admission officer first looked at a book to see whether the college I studied was accredited. Then he looked at my BSc transcripts and said that I have the depth, not the breadth. He advised me to do my teacher education program, BA in Education. While Western Washington University set me back by two years and asked me to take the undergraduate courses they expect of teachers in preparation, I was willing to go through the program. Sometimes, it is better to move two steps back to go many forward. Unfortunately, UBC did not see a path for my educational endeavor on their own campus. That is the beauty of the American system of education, which provides opportunities to excel in education if one is willing. There were many Canadians who came to Western Washington University during the summer to take teacher education courses.

One evening, I took the Greyhound bus to Bellingham, WA via Vancouver, BC from Hazelton, BC. I left our two-year-old son at home with my husband. I was a full-time student for two years, and I shared a room with an undergraduate. Twice, we had fun cooking in the room. That year one of my classmates drove me to her home in Seattle during the Thanksgiving weekend. To this day, every Thanksgiving, I pleasantly remember the extravagant meal and the grand celebration with my classmate's family and friends.

My husband and son called me once a day. I went home for short holidays to be with the family, and our son would say at the Terrace, BC airport, "Go back to Bellingham." But my family was with me during the summer where I was taking classes. Most often, our son would accompany me to classes and the library. Professors valued his presence. He was thrilled to drop the library books in the slot. When someone brings a child to my class, I am happy because I exposed our son to university life early on. He was growing up seeing me read academic articles while still in my arms.

I took two years of undergraduate work in general studies to meet the BA in Education requirements. Therefore, to hasten the process, I did some courses through correspondence and others face-to-face to complete the two-year course within a year, including the Christmas vacation.

As part of my two-year general studies, I took a course on creativity. To demonstrate my creativity, I decorated a cake with hand-crafted sugar flowers that I learned in Sri Lanka during a summer vacation. The professor praised my work. But the grade he gave me for the course was a B, which was not welcoming. So I told him to reconsider my grade. He doubted that I did that creative work. So he told me to write a six-page essay. He was not still convinced. I told him I will show him how I craft sugar flowers. In fact, I noticed the display of the yellow rose, wrapped in clear plastic, on his office wall. At the end of the meeting, the professor said that he will change my grade to A. I was hopeful. But it did not happen. So I went back the third time and challenged his decision. I told him I will take this issue to higher authorities if he did not change it. The next day my grade was changed from B to A. I did not feel right about how I was treated. A grade between a B and A matters when one applies for scholarship funds.

At Western Washington University, I did well in the two-year of course work, including the sciences and humanities. However, our education director at the university did not believe in my abilities. He even challenged me to take a written exam in a big lecture hall with no supervision. He did not say anything after he read my essay. One day, I went to the English department and approached a professor to give me a grade for an English course. I showed my Lady Doak College records, and based on that, he was willing to provide me with a B for a course in English that the college of education would accept. When this drama was going on, I was taking a course on argumentation, which the education department head did not recognize. I did not know what lay ahead when doing a class on argumentation. I argued for a specific structure of the crib that is safe for a baby in an oral presentation.

I did my student teaching in general chemistry, advanced chemistry, and physics with David Tucker at Mount Baker school district, at least an hour drive from the university. I caught a ride with another student teacher who was placed in the same school to teach social studies. I achieved beyond the expectations of the cooperating teacher and the faculty advisor. The latter had told me during my two-week observation, "How are you going to teach our American students?" I pondered about the various remarks people have made about me. But I knew my strength and abilities to do anything that I desire, including piloting Air Lanka between Sri Lanka and India for 15 minutes. It was one of proving myself at every juncture. This is the lot of being in a new country, trying to have a footing, whether it was Canada or the United States.

My cooperating teacher once challenged his physics students to help him figure out a problem dealing with two similar triangles. He asked each student to note the small triangle within a large one. Finally, he asked me. I stated that I see the proportionality of the two similar triangles and began reciting the mathematical expression. This is to that … that I had learned in my grade eight geometry class in Sri Lanka. The teacher completed the statement. A smile of surprise spread on his face with a sense of awe. At the end of the term, he said that he appreciated me taking his advanced chemistry students to the second year of college chemistry. He added that most students opted for examination questions in the units I taught. One thing I remember my cooperating teacher telling me is that I was good with advanced chemistry students. But I need to remind the general chemistry students as soon as they come to settle down, to take their notebooks, take their pens or pencils, and get ready to write notes.

My faculty advisor also gave an English grammar test to the six students he supervised at the end of the practicum. And I got the highest mark. He was wondering why the Americans did not do well. I completed the internship with a very high grade. The first to come out of his office with a genuine smile to congratulate me was the education director who did not trust me or have much hope. I can still picture this happy scenario. Some of my professors in Bellingham are no longer alive, but I appreciate their subtle influences.

2.6 HIGH SCHOOL TEACHING REJECTIONS

In my fourth year at the UBC, I applied to various universities in the United States and two high schools in British Columbia. My advisor encouraged me to apply only to a research university – at the time, I did not know the difference between a teaching university and a research university.

I applied to the two high schools primarily because I wanted to experiment with alternative ways of teaching and learning.

One high school in rural British Columbia advertised for a chemistry position. I applied for the job and was called for an interview. I traveled by bus to the school for more than four hours. I had a heart-to-heart conversation with the principal and superintendent. It did not seem to be an interview. That night, when I returned home, I already had a letter stating that I was not offered the job. While there, the administrators told me that they have not even finished a master's degree. One of the people who wrote a letter of support on my behalf at the UBC also said that she told them not to take me because she felt my place is in a research university.

Next, I applied to a vibrant high school in British Properties in West Vancouver, British Columbia, for a physics teaching job. My minor was in physics, and I had done student teaching in physics. I was called for the interview. One question the principal and the vice-principal asked was what will I do if a student does not submit his or her work on the due date. I said that I will give this student more time to complete the assignment. After all, I have learned at the UBC through example to be empathetic to learners. Now it was time to put that into practice. Both administrators showed me the door. I was not hurt over such actions because I knew that my target was to teach at the university level. I remember my advisor and a committee member at the admission interview asking me what I wanted to do after completing my doctorate. I was hesitant to tell them about my goal. So, they answered the question they posed. This is the kind of empathy that my mentors showed me throughout the program. Now I need to pass on this character to those who I am teaching so they may be transformed.

2.7 UNSETTLING MOMENTS— PhD VERSUS EdD

The April 1991 convocation ceremony was grand. Nevertheless, the thought of receiving an EdD instead of a PhD was troubling. The following year the UBC was offering a degree in PhD in Curriculum and Instruction, and I was prepared to stay for one more year. One of the committee members stated that it does not matter because it is your productivity that counts ultimately. Cedric Linder, who was my peer mentor at UBC, told me that EdD is more valuable in South Africa. After all, Harvard University had only an EdD program for a long time. It is in recent years that Harvard began to offer PhD in Education. At the time, UBC did not have a PhD program. After I left, the program I followed was the new PhD program in Curriculum and Instruction.

2.8 WHAT AFTER ORAL DEFENSE

I wasted no time after my oral defense. I wrote two articles from my dissertation and submitted both to *Science Education*. In the first article, *Chemistry students' conceptions of solubility*, I named my advisor, Gaalen Erickson, as co-author. In the second article, *Relational conceptual change in solution chemistry*, James Gaskell, a committee member, was the co-author. I did not tell either of them before I submitted the articles. I believed that co-authorship was a way of appreciating their involvement in my intellectual growth. After a prolonged review process of the articles, I received word that the editor accepted both articles with minor revisions.

I mentioned the good news to both my co-authors. They were surprised. They both expressed their willingness to help me with the reviews. One of them said that I was very generous. When I submitted the revisions, I was warned by my co-authors not to be too hopeful as I am always. But I knew that both articles will be accepted. In fact, the *Science Education* editor, Richard Duschl, was compassionate. He offered me secretarial help to publish the articles as soon as possible. Because of individuals like Richard Duschl in my academic life, I gained much confidence in research and publication. Although the release of these two articles got switched, it gave me the impetus to move forward. Soon I became a member of the editorial board in *Science Education*. The *Journal of Research in Science Teaching* recruited me to be a member of the editorial board. These were exciting moments and new adventures in my academic life.

2.9 PEERING INTO THE NEXT DECADE

At an interview in the University of Windsor, the question was whether I think science is a noun or verb. I said it is both and explained what I meant by each. I enjoyed my two-seater car ride with the dean of education. However, I did not get the position. I came to understand that the University of Windsor was looking for a teacher–teacher rather than a researcher–teacher.

From Windsor, I looked across the Detroit River. I saw the GM tower, and the place pleased me. I wished someday that I would cross the Detroit River and teach at a university in that city. Ten years later, my wish came to pass. I lived in luxury in the Millender Center in the penthouse on the 33rd floor. There was a beautiful view of the Detroit River. I also had a stunning view of the fireworks on Independence Day. The apartment was connected to the GM building with a concourse. GM has all the facilities. It was one of the city's major attractions. I went for walks in the river walk each day. My relatives who came to see me said that I am leading a hotel life and I am in the center of the city's attractions. I am on the 20th year at the Wayne State University.

I had the rare opportunity to take art lessons from a street artist in the GM River Walk, who was drawing portraits with a lead pencil. In the last few years, I have been taking online art instruction weekly with Darrel Tank through a follow-me-method. He teaches five-pencil, straight-edge, and divider method, which requires scientific thinking, in-depth analytical skill, detailed work, and much patience. This development transfers into science education research and writing.

In my search on the internet, Darrel is the best art instructor in my opinion. I can now produce photo-realistic portraits of people or subjects like honey bee, dog, cat, etc., which takes advantage of knowledge of the anatomy of the subject. Again, drawing realistic drawings require to be "lazy." It appears that I have been engaged in "lazy" and creative activities in life, whether writing, drawing, sugar crafting, scientific dressmaking, scientific cooking, and playing piano and violin. I have yet to learn wood carving, glassblowing, and pottery, which need a "lazy outlook".

Evolving Educator

3

3.1 ACCEPTING AN ACADEMIC POSITION

Shortly after I had graduated in April 1991, I applied to the University of Manitoba, Manitoba, Canada. I was called for a two-day interview. I engaged in conversation with Professor Hal Grunau, (a biology educator), Professor Art Stinner (a physics educator), and their wives over dinner at Grunau's residence the night before the formal meeting. The next day was my talk on research on students' conception. That afternoon, I went for a stroll with Hal along the Red River in Assiniboine Park. This is a place where tourists gather and people ice-skate and play ice-hockey on the frozen Red River. The treatment was royal with mushroom-cheese melt for lunch. There was no doubt in my mind about an offer.

First and foremost, Hal said that I come from a "good shop." Second, my curriculum vitae was impressive, with four articles accepted in scholarly journals such as *the Journal of Research in Science Teaching* and *Science Education*. Third, the Canadian Science Education community was small, and the position was only open to Canadian citizens. So I knew I was in a favorable position. Soon the dean of the Faculty of Education offered me the job with a reasonable starting salary. I accepted the position with gratitude.

3.2 PRESERVICE TEACHERS LEARNING TO TEACH SCIENCE: A CONCEPTUAL CHANGE INQUIRY

The classes began. I was given a reduced load for the first two years to teach two sections of elementary science methods. After the first few years, when the chemistry education professor retired, I taught chemistry methods course. I also taught a graduate course called The Study of Teaching.

I represented the Faculty of Education at the Manitoba Education Department committees to develop the grade ten science and chemistry curricula. The science curriculum director at the ministry was new. The diverse committee members did not pay much attention to what I was suggesting. Constructivist pedagogy sounded strange to the members. It was only years later the science director told me that he understood what I was implying. This was after he became more aware of the current literature and his extensive travel to other provinces of what they were doing. However, a rural chemistry teacher who was a member of the committee liked what I had to offer. While he expressed this privately, he did not come out and say anything at the Manitoba Department of Education (MDE) curriculum meetings about my contributions. Sometimes the chemistry professor at the University of Manitoba and I were at odds. We both were university representatives at the MDE for several years.

I taught preservice teachers without a textbook. At the end of my first year, students said that I should write a book so that they have something to read and follow the model I was advocating. I listened to them. To begin with, I used pertinent papers on teaching and learning science. At the same time, I modeled to them a conceptual change inquiry learning using a unit on light. For my own resources, I followed Feher and Myer's (1986) articles on light and shadow in *Science and Children* and *Science Education*. The exploration activity on light was discrepant. I also had a copy of a unit on light from a conceptual change perspective authored by a group of science educators in New Zealand. Both materials were research-based, and I was attempting to translate research into practice. Karen Myer came to the University of British Columbia (UBC) to do her doctoral studies a year later. Karen observed one of my classes at the UBC. At the end of her observation, she said that I was truly a "practicing constructivist." By that, she meant that while most science educators talk about constructivism in class, I am actually modeling it for elementary preservice teachers and helping them make sense of what I am doing.

A conceptual change model of teaching and learning consisted of eliciting children's ideas with an activity – a task or cards with pictures to represent the phenomenon. Children's ideas were grouped into themes, and were interpreted as everyday ideas, naïve ideas, or misconceptions. Then, students were "confronted" with activities that would provide them sophisticated scientific explanations. At the time, the best conceptual change model consisted of four phases developed by Posner's group in 1982: (1) An idea manifested crisis; (2) a plausible or likely idea was put forward; (3) the plausible idea was considered intelligible by the learner. The naïve notion was replaced; (4) the accepted idea became fruitful when applied in several cases. This conceptual change model stemmed from the development of scientific knowledge. Other conceptual models were grounded in Piaget's theory of human development – assimilation and accommodation. Assimilation was considered when the idea was partially correct and required clarification. Hence, another idea was added to the naïve notion, or the naïve idea was elaborated. Accommodation meant that the original idea was rejected and replaced with the scientific concept. This model was founded in the development of human knowledge. While the conceptual change models differed philosophically, the teaching was similar. Elicitation of children's ides was common to both philosophies. Similarly, clarifying and/or replacing the naïve idea is the same in both philosophical camps.

The preservice teachers in my class were not introduced to the philosophy or the theories of learning. They read the collection of papers for a better understanding of my modeling. Preservice teachers were expected to make sense of what I was doing. They were asked to develop a short unit plan using a concept from the Manitoba curriculum guide with the ideas taught. The students could not reconcile the way I taught them to teach elementary science. This was never their experience. Their experience stems from their school and university science learning. The model was contradictory to their ideas of teaching science. The preservice teachers themselves were placed in a learning environment that enabled them to experience a conceptual change about teaching and learning.

While eagerness to translate theory into practice was sensible, it also became a rift between the elementary preservice teachers and me. They wanted me to engage them in simple activities taken from school science-related activity books. Their belief was they should have a collection of ready-to-go, hands-on activities to use in their practicum classroom. My focus was on developing preservice thinking about teaching and learning. In fact, they were learning to teach based on a defensible framework. Preservice teachers' focus was on themselves, that is, learning to present science through activities. I expected them to focus on students' learning – this concept was completely foreign to them. The focus was on self and attempting to best present science concepts through follow-me-demonstrations and scripted laboratory learning.

In my elementary methods course, I conducted a Preservice Teacher-As-Researcher (P-STAR) conference. Each student presented a paper about his/her learning to a small group of students. This presentation simulated poster presentations at an education conference. Each student also had to submit a short paper. The performance was graded by peers based on a simple rubric we had developed while I graded their paper. One presentation stuck in my head. We were getting ready for the P-STAR conference. A male preservice teacher asked me whether he can present his paper on conceptual change from a Biblical perspective. He told me about his church background. We discussed what he was going to present.

There were five presentations and students had the choice of going to the presentations that they were interested in. I was pleasantly surprised to see a big group gathered around the conceptual change presentation. That entire semester I had modeled conceptual change teaching and learning from both psychological and scientific perspectives with inquiry activities. And the students found the approaches challenging.

3.3 THE WOMAN AT THE WELL CONCEPTUAL CHANGE

The Christian preservice teacher spoke on conceptual change from a Biblical viewpoint. After his peers listened to the presentation, they acknowledged to me that they now understand what I was talking about. I also realized that students I was teaching are Christians and Manitoba is a Christian belt. These students perceived science as "static," and the science curriculum provided by the Department of Education in Manitoba is hard and fast. Every detail in the curriculum should be followed as the letter of the law. The students were not happy with what I discussed in class because I was going against their beliefs. This was a recurring problem during the ten years I worked in Manitoba.

My student's analysis of the Samaritan Woman at the Well allowed me to think about a long-term project. I wish to write a book entitled: Jesus' Geniuses: Science Pedagogy of Love and Compassion. A passage that I wrote on conceptual change is described below.

Teacher–Student Discourse in the Learner Space:
The Tension between Theoretical and Real

John 4: 4-15 New International Version

After a hard morning's work, Jesus rested at Jacob's Well, near the Samaritan town of Sychar. At the same time, his disciples went into town to buy food (6). It is a well-known custom of women to fetch water together at the Well in the early part of the day. It is a highpoint of a women's life when they would meet and exchange news with one another. But one woman came alone in the mid-afternoon – she was considered an outcast among women in her community. It was customary for women of ill repute to fetch water from the Well at noon, to avoid meeting others in their community.

Jesus asked the woman who came to fetch water to give him a drink (7). The woman replied, "How can you, a Jew, ask me, a Samaritan woman, for a drink?" (9) This discourse with the Samaritan woman reveals that Jesus was crazy breaking three Jewish customs. First, he initiated the talk with a woman that highlights both race and gender issues. It was customary to not speak with women, and the disciples had a demeaning view of women. But Jesus broke the social rules because he reached out to a woman. Second, she was not only Jewish but is even ostracized within her own Samaritan community. Third, Jesus asked water from her, which makes him ceremonially unclean because of using her utensils. According to Higgs, L.C. (2008), Jesus broke the social rules because he reached out to the woman.

Jesus drew the woman into a deep conversation. He made extraordinary spiritual statements drawn from her own context. "If you knew the gift of God and who is saying to you, 'Give me a drink,' you would have asked him, and he would have given you living water" (10). The words Jesus uttered create in her a sense of awe and wonder. The woman probes Jesus about her sensual reality of the bucket and the depth of the Well (11). She talks about the greatness of the forefather and his children who used the Well (12) She confuses her physical and spiritual needs. Salvation is far from her thoughts. Yet Jesus gently leads her along to realize her spiritual needs by giving a spiritual answer. "Everyone who drinks this water will be thirsty again. But whoever drinks the water I shall give will never thirst. The water I shall give will become in him a spring of water welling up to eternal life." (13, 14). She was eager to fulfill her physical needs but overlooked what she lacked spiritually.

Intrigued, the woman asked him for that "water of life" so that she would no longer need to draw water from the Well. Jesus assures her that she needs this 'water' and is better than Jacob, her forefather who had dug this well. She wants Jesus to meet her human needs (15). There is tension between Jesus' spiritual statements and the physical meanings she attaches.

More than 30 years of conceptual change science education studies clearly reveal that there is tension between students' ideas of science concepts and scientific explanations. It is not easy to move them from their personal frame of reference to scientific explanation. When scientific explanations are given, students understand in light of their reality rather than the scientific theory.

3.4 RESEARCH IN THE ELEMENTARY SCIENCE METHODS COURSE

I competed for the Faculty of Education Research and Program Development Funds of $3,000 per year over two years (1992–1994). *Preservice teachers' conceptualizations of and attitudes toward science pedagogical content knowledge* was a collaborative project conducted during 1992 and 1993. *Preservice teachers' conceptions and their science instructional practices* took place between 1993 and 1994. I also received $4,202 for the project: *Preservice teachers' meaning-making of science instructional practice: Collaboration and construction* from the *University of Manitoba/Social Sciences and Humanities Research Council of Canada* Grant.

With the above funded projects, I was able to write two articles based on my experience in methods course. The first article based on preservice teachers' reflective papers was published with Adel Hay, a master's student at the time. *Preservice teachers' meaning-making of science instruction: A case study in Manitoba* was published in *the International Journal of Science Education* (IJSE) in 1995. In this paper, I had included how a preservice teacher viewed science instruction from a Christian perspective.

> As a Christian, I believe that God has created the world and given mankind the ability to discover the truths which He has hidden in it. He gave us minds to exercise and imagination to use. Those are abilities we must not deny our children, either. As a teacher, I believe I must encourage my students to ask questions about the world around them, to seek new ways of finding answers to their questions, and to wisely use the knowledge they gain from their enquiries. While this seems to fit with parts of the constructivist approach to teaching science, I have a difficult time meshing certain aspects of the approach with my personal beliefs. For example, I believe that there are truths which God has built into the universe both scientific and theological. Because I believe that this is so, I also believe that children should be encouraged to discover those truths. Here is where I find a dilemma with constructivism. Constructivists would say, I think, that people construct their own knowledge based on both their experiences and their understandings of those experiences. This seems to say that there no universal truth to be discovered, but that truth (scientific or otherwise) is relative to the person considering it. I cannot make logical connection between these two ideas – that truth is universal and created to be discovered, and that truth is relative (Ruth).

(Ebenezer & Hay, 1995, p. 103)

This excerpt piqued the interest of the editor of the IJSE. He invited me to explore the Christian preservice teacher's quote further. Also, he asked me to interview preservice teachers of different religious groups. However, I could not do the latter request because the composition of the class was whites. So I hired a student from the same class using the grant funds that I received and asked him to interview his classmates. The audio-recorded individual interviews were transcribed. During this time, I attended Fenstermacher's seminar on practical arguments at the University of Manitoba. His work inspired me to look at the practical arguments that Christian preservice teachers offered in the science curriculum and instruction course. Consequently, *Christian preservice teachers' practical arguments in a science curriculum and instruction course* was published in *Science Education*. I was pleased that the associate editor called me and thanked me for the scholarly article on the practical arguments of Christian preservice teachers.

3.5 EXCEPTIONAL STUDENTS

Some preservice teachers performed beyond the elementary methods course. I narrate my experience with two such teachers.

3.5.1 Jennifer

Jennifer was in her early twenties when she was in my science methods course. At the end of the semester, she said, "I enjoyed your ideas of teaching. Would you help me with how to teach with your ideas in my practicum?" I told her that I am not her faculty advisor and I cannot walk into her student teaching classroom. If she really wants me to, I will help her teach while doing research with her about her learning to teach science. I asked her to choose a concept for her grade two classrooms, and she chose flotation. I told her to prepare an elicitation activity, which she did. She explored grade two students' ideas on flotation. She took the students' ideas and sketches on the concept of flotation home and grouped them like I had taught in class. We decided that she will represent students' ideas in a concept map so that she can expose their ideas to them. It was a straightforward concept map, which she constructed on a Bristol board and hung in her classroom throughout the unit of study.

She pointed her students to the concept map and said that they were their ideas and that she is going to use their ideas to teach them about flotation. She asked the students to bring objects from home to do a group activity. She gave each group an activity that would allow them to arrange the objects into two categories, objects that would float and those that would sink. The grade two students gave her reasons why some objects floated while others sank. In this unit of study, students brought objects to every class. One student even brought a life-sized tube, and another brought a doll with flotation devices and experimented with water.

The principal walked in several times into this classroom. He did not say much, but I was wondering what was going through his mind. The class was noisy with playful science. My mind went back to my own teaching of grades eight to ten science in a school. The teachers and the principal complained that there was too much noise from my science classroom. At that time, I did inquiry teaching with no theory attached to it.

After the unit on flotation was over, I left the school. At the end of the term, the principal hired Jenifer to teach grade four in that school. I received a letter from Jennifer during Christmas. She brought my attention to two things. One, she was taking baby steps in enacting the ideas I had taught her. Two, the principal had asked her to write her reflections and called her to discuss with him. Jennifer said that her principal remarked that her thoughts point to her as a ten-year experienced teacher.

My proposal based on our research on grade two students' learning in a unit on flotation in Jennifer's practicum class was accepted by the XXIII CSSE Annual Conference, Montreal, Quebec. The paper was on *Post-modern science education: a joint exploration* at the XXIII CSSE Annual conference, Montreal, Quebec. Interestingly, Jennifer, a Catholic by faith, invited me to pray before we presented the paper. Jennifer offered a short prayer before we walked into the presentation room. She also prayed for our conference presentation before we started our journey to Montreal from Winnipeg. Being a Christian science educator and working in a public university, I did not think it was appropriate to pray in general settings. However, a student's background was valued and cherished. Currently, in Michigan, one of the high-level practices in teaching is to accommodate family, culture, religion, etc. I suppose praying with Jennifer at her request is one way of respecting a student's religious background.

The educators who had gathered in the room applauded Jennifer's work. They stopped to talk to her, and I was so proud of her. When we went back, the student body that sponsored Jennifer to the conference stated that she should, upon her return, give another paper presentation. So I organized this event in my own graduate class, the Study of Teaching and Learning. Jennifer took the entire three-hour class. The graduate students were amazed at the knowledge that Jennifer portrayed. The graduate students said that even they, as experienced educators, did not have an in-depth knowledge and understanding about teaching and learning that Jennifer had. I pointed out to the students that this awareness is an outcome when a teacher engages in research about her teaching and student learning. Besides, my belief and admonition that teacher-as-researcher is critical to meet accreditation standards. One standard requires evidence of preservice teachers' monitoring their practicum students' conceptual growth.

After a few years, a different Jennifer invited me to observe her class to see whether what she was doing aligned with what I taught in the methods course. I had not asked permission from the school and had expected Jennifer to do this. Lo and behold, the principal came out of his office to meet me at the school entrance. He took me to his office. I was concerned. The principal asked me whether I recognize him. I said, "No." He said, "He is the same principal who was in the school where I went to help Jennifer with her grade two unit on flotation that she enacted during the practicum." Then he added, "My nephew loves his teacher, and this is that Jennifer." Finally, I did not spend much time with the second Jennifer, who asked me to observe her grade six, science class.

A few years had passed. Sarah, a preservice teacher called me to help her. A teacher with 30 years of experience was her cooperating teacher. The grade eight students seemed very bright. I was impressed with the conversation that they had with me about the science concepts that they learned in class. They provided evidence and reasons for the claims they made. Such learning was inherent in them. I wished I had a tape recorder to record what they stated. I lost a golden opportunity to document their ideas based on an entire unit of study on matter.

I talked to the classroom teacher about the students at almost the end of the practicum. She then told me, "Don't you recognize these kids." I said, "No." She then told me, "They were the same students that you helped Jennifer teach a unit on flotation in their grade two class. And I have taught for 30 years, but this is the best group of students I have ever taught," and she thanked me for providing the foundation of science learning. I wonder what Jennifer and her grade two students and the same grade eight students are doing now?

3.5.2 Anna

After the Curriculum and Instruction course was over, Anna called me at home one evening and announced that she got a position in an elementary school to teach grade six. She told me that the only question the administrators asked was whether she knew HyperCard, to which Anna said, "Yes." I was excited for her and jokingly asked whether she knows how to spell HyperCard. It was customary that if I learned something new I would introduce that to my students. I shared some HyperCard activities with the elementary preservice teachers in the methods course. They were awed by the animations I had created. However, scripting is not tricky. Grade six students can learn the program very quickly. I did not doubt in my mind that Anna would grasp it within one day.

In October, I unexpectedly saw Anna and her 24 grade six students at the Manitoba Technological Center. They had come to see the exhibition on technologies, while I had gone to see the exhibits. After a few years at the end of the year, Anna came to my office. She excitedly told me that she applied for a position in Toronto, and she has an interview. The reason she came to see me was to ask for the article that she and I had co-authored. *Fostering common knowledge through problem-solving* was published in *The Manitoba Science Teacher* in 1994. I had only one copy of the journal, which carried the article. Although she promised to mail it back, she never did. She did get the position, and I was wondering why Anna? Aren't their teachers in Toronto?

3.6 OTHER PUBLICATIONS

I have published articles with other elementary preservice teachers as well in the Manitoba Science Teacher. All these articles were about classroom practices. Lisa co-authored *What do children really think of shadows?: A preservice teacher's explorations of science instruction* in 1992. Carla was the lead author on *The POE strategy for chemical knowledge construction*. At the end of the year, Carla was selected as the best student of the year. She, in turn, chose me as the best teacher of the year for the Outstanding Students Honoring Outstanding Teachers Award at the University of Manitoba. There was a celebration and my name is etched on the wall outstanding teachers.

The *Manitoba Science Teacher Journal* editor has often requested me to submit articles. In 1996, I sent two articles. *Developing common knowledge: A perspective of measuring success in science class* was reprinted in *Catalyst*, British Columbia, Canada. Another article was on *Environmental ethics in sustainable development education*.

3.7 TWO YOUNG MALE PRESERVICE TEACHERS

I challenged the elementary preservice teachers to use Desktop Publishing to publish their science units. Two male teachers responded to my call. The following year, they came to my office to thank me for giving them unique opportunities. They mentioned that because of the experience they had with developing units, they were hired by school districts to become technology coordinators with a higher salary. Both teachers had expressed to me their desire of eventually becoming principals of elementary schools.

An interesting phenomenon was while I was still in Manitoba, American superintendents called me several times. They thanked me for preparing great teachers, and were pleased to hire my students.

3.8 WRITING TEXTBOOKS

One morning, a representative from MacMillan Publishing Company made a surprise visit to my office. He wanted me to take a look at the elementary science methods course books he brought. First, he asked me whether I use any textbooks. I said, "No." I told him that I don't want to use any books because I have my ideas to prepare preservice teachers to teach science. He was curious to know my thoughts. We talked for a long time. I told him about the compilation of my papers and requested him to get it from the bookstore. The next morning, I received a call from Toronto, telling me that they are going to send the book to their counterpart in Ohio, and I will hear from them. Linda Montgomery, the editor, called me and discussed the book and stated that they are praying for a book that has content in the compilation of papers. They asked me to sign a contract for two books, one for elementary and one for secondary.

The editor stated that she is going to send the spiral-bound copy of the collation of my papers for review. Of course, as expected, the compilation did not get a useful review because it was not a textbook at that point. Anyway, they asked me to start writing books. I co-authored the elementary book *Learning to teach science: A model for the 21 century* with Sylvia Connor in 1998. In 1999, the Canadian edition was released. In 1999, I co-authored the secondary book *Becoming Secondary School Science Teachers: Preservice Teachers as Researchers* with Sharon Haggarty, a UBC graduate, who was always looking out for me as a senior professor at the University of Western Ontario, London, Ontario, Canada. While writing the textbooks, Linda Montgomery came from Ohio to observe my classes. It was during the P-STAR conference. She appreciated what was happening in class and how I prepared the teachers.

In total, 18 reviewers, all from the United States, reviewed the books over three years, and wrote positive reviews. One reviewer mentioned that the company should distribute the book to 45 research institutions in the United States. At the time, the contents of the books were novel and radical. The science teacher educators who wished to use the book should have a strong foundation in theory. The foundation of the book was conceptual change inquiry from a relational perspective. There were several examples of my own preservice teachers' work. In the face of reform and national education standards currently, the two books would be useful in science methods courses.

3.9 TWO ACADEMIC FAITHFUL

3.9.1 Cedric Linder

Cedric Linder is a South African from the University of the Western Cape (UWC), Cape Town and is currently Chair Professor of the Physics Education Research program at Uppsala University in Sweden. Cedric had already been at the UBC for a little over two years when I started my doctoral studies there and was soon to start writing his dissertation.

Cedric was very welcoming and supportive. For example, he was always willing to read my paper drafts and give encouraging feedback. When I got "writers block" on some aspects, Cedric would ask me to explain to him what I wanted to say. Then he urged me to go and write it just as I had explained it. And it worked!

As our collegiality developed, our families would get together for dinners, and these times together soon made us good friends. After Cedric successfully defended his doctoral dissertation, he went back to the UWC where he taught physics. We kept up our friendship and collegiality through email correspondence. Later, Cedric arranged for a younger colleague, Delia Marshall, to visit me at the University of Manitoba. This later led to Delia and Cedric inviting me to South Africa as a visiting professor. While at UWC, I provided input for the future growth of the UWC collaborative research program that Cedric and Delia were building. This led to them inviting me to be the discussant for a paper on girls in science at the "Public understanding of S & T" conference that they helped host in 1996.

While visiting UWC, Cedric arranged for me to give a presentation to the entire Faculty of Science and Mathematics on multimedia for teaching and learning. I also gained experience helping supervise some of their doctoral students and acted as an external examiner for one of their master's students in the area of Physics Education research.

Joan Solomon from Oxford University had also presented a paper at the "Public understanding of S & T" conference. Before coming to the meeting,

I had been contemplating spending a sabbatical at Oxford, so I sent a note to Joan. I also told Sylvia Connor about this. Sylvia, who had retired, had been very supportive of my faculty application to the University of Manitoba. At the time I was busy taking her place, and we collaborated in writing two books – see the discussion below. Sylvia asked me to phone her. She was seated in my office while I called her and on the other end was Joan Solomon. Unfortunately, Joan was initially not very supportive of my idea to spend a sabbatical at Oxford. However, after receiving my CV, Joan was full of praise for the productivity that I had achieved within five years of receiving my doctorate and she invited me to do collaborative work with her. Thus, I started to seriously contemplate going to Oxford for my first sabbatical. However, after discussions while at the "Public understanding of S & T" conference, I started to feel less enthusiastic about spending a sabbatical at Oxford. Instead I did two things. First, I spent about half my sabbatical time at a private university in Bangladesh where I helped develop a teacher education program. Second, I spent the remainder of my sabbatical time with Richard Duschl at Peabody College, Vanderbilt University.

While visiting Cedric at UWC, he introduced me to Duncan Fraser, a Professor of Chemical Engineering who had helped set up the Centre for Research in Engineering Education at the University of Cape Town (UCT). Cedric and Duncan were very close friends and had a very strong and productive collaborative research program together until Duncan passed away a few years after his retirement. Duncan, and his wife, Jenny, a medical Doctor, took me to their cell ministry, which involved visiting a poor family in a one-room tin shed covered with newspaper. There was one bed and a tiny space for their kitchen with a stove and a few utensils. The hostess served us cookies and tea, perhaps specially bought for us. This amazing and rewarding experience was very special. It was inspiring to witness a chemical engineer from a prestigious university in Cape Town, and his wife, a family Doctor, reaching out to the less fortunate in this way. This experience is a demonstration of how much more teacher educators and those who teach science should reach out across social and economical boundaries to make real difference in people's lives.

One day, Cedric took me to Duncan's office. By this time, Cedric had moved to Uppsala University keeping in touch with his UWC Ph.D. students with regular visits to Cape Town. Duncan asked me to join him in some research into students' understanding in the area of thermodynamics. We decided to conduct a phenomenographic study with his first-year chemical engineering students. Our collaboration went well because I could directly draw on my doctoral work for the project. This led to Duncan posing if it would be possible for me to spend six months working with him at UCT. I was unable to accept a six-month-long invited sojourn at UCT; however, I did continue to work with Duncan for the remainder of my time in Cape Town. I set up three activities in chemical energy to elicit 21 first-year engineering students' ideas of chemical energy. I then suggested to Duncan that he use our results to help inform the new curriculum design that he was currently involved in. New and radical ideas emerged from our discussions, and Duncan said that he felt very nervous about introducing them into the curriculum at UCT. Our collaborative work continued and Duncan and I started to get our work published. First, there was the variation in ways students conceptualized energy, and then, with assistance from Xiufeng Liu, we published a paper on dealing with the structural characteristics of university engineering students' conceptions of energy.

On the day of my departure, Cedric took me to the airport, and while waiting for the call to go to the gate, we had a cup of tea together. At this point, Cedric gave me a card asking me not to open it until I got onto the plane. The message in the card was about how much he had come to value our friendship. That card meant a great deal to me. Over the years that followed, Cedric continued to support my work, for example, having me become an assessor for the South African National Research Foundation, which I found both challenging and rewarding. Since then I have been frequently asked to evaluate promotion, grant, and research chair applications by South African institutions.

3.9.2 Xiufeng Liu

Within the first few years, I became a member of the coveted doctoral studies committee in the Faculty of Education at the University of Manitoba because of my rich research background. I also received a multiyear grant from the prestigious SSHRCC in 1994. I was also the co-PI in Xiufeng Liu's SSHRCC grant.

Xiufeng and I have shared the same office at UBC. Since then we have worked together, he was at St. Xaviers., Nova Scotia, Canada. In 2002, Xiufeng and I published an article in *Research in Science and Technological Education* on the *Descriptive categories and structural characteristics of students' conceptions: An exploration of relations.*

I was responsible for Xiufeng to find his way to the United States. Xiufeng is a star in science education. I have included Xiufeng in all my NSF proposals in some form, doctoral committees, and wherever I can use him for my students and my benefit. Xiufeng has contributed much to the national and international science education community and he is a world-renowned science educator.

3.10 IMPRESSIONS OF TWO SCIENCE EDUCATORS

3.10.1 Arthur Stinner

Arthur Stinner was a colleague at the University of Manitoba. In his physics methods course, he taught from a historical perspective. An assignment that Art gave his students was to write a paper on the history of light, motion, electricity, etc. In their essays, students included the experiments that early physicists carried out, as well as the influence of the cultural, political, and religious contexts on scientific knowledge development. The paper also showed how early physicists' work can be translated into a lesson plan. His orientation to science teaching appealed to me. Hence, I read several papers about the history of solution chemistry. The readings were fascinating. I discovered that there was a parallel between historical ideas and students' ideas. I constructed a chart showing the similarity. See Table 3.1. I share this table with education students who take my methods courses.

During the 1999–2000 academic year, I competed for the University Research Grant to conduct *A historical analysis and interpretation of the chemical theory of solution*. At the same time, I was keen on Michael Matthews' work on the history and philosophy of science. I expressed my interest to attend one of his conferences. So, he invited me to give a plenary speech and a workshop in the Third Annual Conference on the History, Philosophy & New South Wales Science Teaching at The University of New South Wales. My talk was entitled *History of chemistry and conceptual change*. The workshop was on *Students' and chemists' early views of solutions: Implications for teaching*.

Introducing preservice teachers to a historical perspective is critical for several reasons. Preservice teachers can readily discern scientific knowledge growth and contextual influences. They can also view parallels between early scientists' ideas and their personal ideas. Additionally, they can use similar experimental tasks to teach their students. Moreover, learners will come to realize that their ideas are related to early scientists, and test their ideas just as the early chemists. Students' ideas can be the basis for argumentation. Their developing and developed ideas based on evidence through tests can also be the basis for argumentation. This is why I consider students' conceptions from the nature of science viewpoint and convey that message to students in the methods courses, as well as to science educators through my publications.

TABLE 3.1 Relationship between scientific ideas and students' conceptions

SCIENTIFIC IDEAS	STUDENTS' CONCEPTIONS	CATEGORIES
Fourcroy – a solid melting in a liquid and partaking of its liquidity	Water molecules are entering into the salt molecules and making them liquid	Thermochemical theory – Liquefaction or melting theory
Mendeleev – The formation of different hydrate compounds at different temperatures	Salt is reacting with water and forming salty water, and forms one kind of chemical thing $NaCl + H_2O = NaCl \cdot H_2O$ Salt water	Thermochemical theory – Hydrate theory
Hermann Boerhaave – Chemical reaction essentially the same as solution. Berthollet – There was no fundamental difference between solution and chemical combination	When salt is added to water it dissolves in water. For example, if NaCl and water is mixed, NaOH is created and H_2 is evaporated. In dissolving atoms get separated from the orbit. Atoms are running away from their shell and joining and making a new shell	Chemical combination theory
Arrehenius – Certain substances dissolved in water because of their electrical properties. He concluded that the active molecules were ions, the charged particles meaning that potassium in potassium chloride is different from atomic potassium, and would not dramatically react with solvent water. Ionic theory also explained Hittorf and Kohlrausch's results. The properties of a salt are the sum of the properties of ions. Faraday and Volta's work on electricity also helped to prove ionization as a mechanism	Ions are moving in water. Salt dissolves in water and forms Na ions and Cl ions.	Ionic theory

Aligned with the history of science, the study of the nature of science becomes meaningful. Articles by William McComas, Norman Lederman, and others have influenced my thinking. I teach an advanced course to science teachers, which focuses on the nature of science. Teachers are expected to write a paper on the nature of science and develop lesson plans that reflect the nature of science. I have worked with Ling Liang, a science educator at La Salle University, and her collaborators on the nature of science and published two papers in 2008 and 2009. *Assessing preservice elementary teachers' views on the nature of scientific knowledge* was published in *Asia-Pacific Forum* on science learning and teaching in 2008. The second article was an international collaborative study published in 2009 in the *International Journal of Science and Mathematics Education*. It was about preservice teachers' views about nature of scientific knowledge development.

3.10.2 Glen Aikenhead

I was impressed with Glen Aikenhead's research on science, technology, and society, a significant strand in science education. He spent his career working at the University of Saskatchewan, Saskatoon, Canada. Based on his work, I began to help preservice teachers understand teaching from an issue-based perspective. This orientation was after reading some of his papers, books, and resource materials. This pedagogical practice is, in fact, more challenging because teachers need to integrate various disciplines, including the need to know science. It was moving back and forth between learning an issue and learning science. This sort of situated or conceptualized science learning had variations: project-based learning, place-based learning, and large context problem-solving. Writing the elementary and secondary textbooks enabled me to dive into various aspects of science education.

Glen Aikenhead gave me one excellent piece of advice. He told me that he used to think working at a larger university such as UBC was usually a goal for most Canadian science educators. He realized the university does not matter. It is what we contribute to science education through our writing that counts. A smaller university is better because there are not too many expectations. Glen Aikenhead was an internationally acclaimed science educator. He encouraged me to remain in one place and do my writing. Another piece of advice he gave me was to leave science educational issues for others to solve and not tackle various topics. Possibly, this is why the integration of technologies into conceptual change inquiry remained my main research focus. On this note, when I became a full professor, our dean Paula Wood read a biosketch about me in the faculty assembly. One striking statement that I recall from what the dean read based on one of the national reviewers' reports was the following statement: "This researcher is a true scientist building the same theme in every investigation she undertook."

3.11 INTERACTION WITH INTERNATIONAL SCHOLARS

3.11.1 Osman Kaya

One day in 2000, Osman sent me an email message stating that he is doing his master's thesis based on my elementary textbook. He said that at Gazi University, Ankara, Turkey, there was no science teacher educator to help him. He asked me whether I would help him write the thesis. The thesis topic was on multiple intelligences after Howard Gardner. I was delighted to help him. However, I was not curious to find out anything personal. It was strictly intellectual email correspondences. After a year or so, he asked me whether I would help him present a paper at a conference. I told him to come to the United States to present his paper at the AERA and NARST conferences. He was delighted to go to the United States to present his research. In the meantime, I moved to Michigan in August 2001. September 11 was the twin tower attack. However, I continued to work with him. He applied both to AERA and NARST, and his papers got accepted for paper presentations. Next, I sent Osman an invitation letter to secure his J-1 visa. Upon arrival, he told me he had to get the official permission from Firat University, which sponsored him to do his masters at Gazi University. Osman applied for his visa, and he got it. My colleagues were reminding me that the United States may not grant him the visa because of 9/11. Osman was able to come on a three-month visa to the United States.

I could not find a place for Osman to live because of his short visit. Furthermore, I know that it is expensive for a foreigner to live in the United States with his funds. I sent a message to my husband in Jamaica, who was a missionary professor teaching business at Northern Carribean University. I told him I am unable to find a place for Osman to stay. My husband had kept up with Osman's academic life through his emails to me. My husband told me, "Lay the bed in your own apartment," at Millender Center downtown Detroit. It was one of the two luxury apartments where GM workers, university students, and other professionals live. I had also given refuge to a nursing student and one of our relatives without any remuneration. I wrote to Osman that he can stay with us and that I am a vegetarian and would not be serving any non-vegetarian foods.

The day before his arrival, I had invited an older Jamaican couple for Sabbath lunch. I was telling excitedly about Osman's arrival on Monday. They both rolled their eyes. The woman asked me whether I know the man who I am going to keep in my home with two other young girls. I said I do not know him at all personally, but we have been writing to each other at least for two years. I told her that my husband knows him through our email messages. She raised a lot of what-if questions, which made me uncomfortable. That night I sent Osman an email asking him, "Who are you?" I asked him whether he smokes. Osman replied and attached his family photo. Osman said that he is 26 years old, married, and has a newborn baby girl. He does not smoke. I was relieved.

I was excited to pick him up from the airport. I told him to be on the curb. He had never seen me before, perhaps pictures on the internet and the textbook he was using. I started talking to him in English. He was not replying to me. We did not know each other's language. I took him home and gave him a big solid air mattress for him to sleep in the living room. The next day I took him by Amtrack to Philadelphia for the AERA conference as I thought it will be cheaper to go by train. Also, I had never traveled on the train in the United States and thought it would be fun. We took a bus to Toledo and had to catch the train. The train was late and reached Washington, DC, only in the afternoon. My cousin came to pick us up. The reason for stopping in DC was to show Osman the nation's capital. We both stayed in another cousin's family arranged by my closest cousin. She had a guest in her own home and could not accommodate us. They were all living in Silver Spring, Maryland. That day, I took Osman to DC on the transit. Even now, I am very nervous about taking the DC Metro because I do not know how to put money in the ticket machine and get the money.

We walked a lot along with Abraham Lincoln's statue to the White House and all other places of interest. We also went to the Smithsonian Institute. While we were walking, I asked him what he was going to present. He said, "Concept Map and Vee Diagram" ever so quickly and I could not understand his diction. We were off to Pennsylvania by train the following day. That whole night I helped him with pronunciation. His presentation was the next day. There was a room full of people. He was quick to learn and his diction was proper. Tears of joy were flowing down my cheeks. People gathered around him to speak with him. I watched this from a distance. He would get up and walk out of other presentations because he did not understand the language. I advised him to stay and listen so that he improves his speaking skills.

On our way back, we stopped in Washington, DC. We stayed at my cousin Merlin's place in Silver Spring, Maryland. Young boys were watching a Hindi Movie that Saturday night. Osman was jumping up and down, enjoying the movie. From a distance, I was wondering what does this fellow know about Indian movies. After the movie was over, I asked him, "How do you understand Indian movies?" He said Indian films are viral in Turkey. My cousin's youngest son and his friends who were watching the movie together told him don't go back with me and stay in Maryland, and they will drive him across the United States. Osman replied, "I came here to study with Dr. Ebenezer. And that's the only thing I will do here." My cousin, in whose house we stayed on our way to Pennsylvania, said, "He respects you a lot." Osman and I went back by Amtrack.

The next day, I introduced him to my students in a graduate class. They all liked him so much. The students were very kind to him. After a week, we went to Chicago to the NARST Conference. I did not coach him this time, but he did an excellent job of presenting his paper.

When we finally settled at home, Osman came to me with his paper on multiple intelligence. He wanted me to help him publish the paper. I told him no science education journal will accept an article on multiple intelligence. But I told him that I will help him to write the article, and that he can submit it to any science education journal. We spent much time with the paper, and the article was rejected. But he did learn how to write and submit an article. The whole time Osman was here, he was not happy when I focused on other people's work. He wanted me to concentrate on his work completely. What he did not fail to do was he would spread a mat and pray to Allah each day. I appreciated his belief and steadfastness in prayer. I told him often please pray for the NSF grant proposal I am writing, and it will be successful. I went to church regularly on Saturday. I took him to church so that he can participate in the Sabbath school discussion and learn to be more conversant in the language. An African American member of the class gave his Bible to Osman. One day Osman asked me if whether I will go to the mosque with him. I said yes, I will go with him. That's when I told him why I am taking him to church. The Sabbath school members liked Osman and the intellectual contribution he made to the class. Within three months, by attending all my classes, going to Sabbath school classes weekly, and accompanying me to all social meetings, Osman became conversant in English.

At home, one of the girls showed much caring toward him. When the food was low in the fridge, she would remind me that Osman does not have any food to eat. These two got along, but the nursing student did not get along with him. Osman had lost 12 pounds staying with me. He also missed his newborn baby and his wife very much. The three-month stay was a long time. The day he left the United States, I was in California.

That summer I was going to Cyprus to present a paper at the conference. On our way back, my family wanted to stop at the neighboring countries. Our first stop was in Turkey. We were going to stay with Osman for two weeks.

Osman came to pick us up at Istanbul airport. As soon as he saw me, he came running and hugged me. I was wondering what the bystanders might think in that culture. Then we all went to the hotel where he had booked us a room. He, his wife, and the child stayed with their relatives. For two days, he took us around Istanbul, including the mosques. Osman also located the Seventh-day Adventist church I attend in Istanbul without me telling him. He came with us. It was a small gathering of foreign nationals. The pastor and his wife were Turkish. At Sabbath dinner, upon the pastor's wife's request, I related how we got to know each other. Upon hearing the story, she said, you have one son. Think of Osman as your second son. I said, "I will."

My husband and I went to Ankara, the capital city of Turkey, where he was now doing his PhD. When Osman came to Detroit, his intention was to share the topic he was thinking for his doctoral thesis in science education. I suggested to him that he should focus on argumentation and the nature of science. I shared some resources with him. Accordingly, his PhD was on the impact of argumentation as a learning strategy on middle school students' view of the nature of science. He introduced me to several scholars and administrators and translated our dialogue. The chemistry education PhD students said that they know me because of my publications. He took us to Cappadocia, a well-known tourist attraction.

While Osman was in Detroit with me, Muammar wrote to me for help with writing articles. I thought Osman must have told Muammar Çalik about our work together. Now that I was in Turkey, I told Osman to give Muammar a call. Muammar, with his newly wedded wife, were on their honeymoon 12-hour bus ride in a coastal area. Immediately, that night, they took a bus to see me in Ankara. We had a good time that day, although language was a significant barrier. That night, they wanted to go back to Karadeniz (Black Sea) where Muammar was teaching chemistry teacher education.

After two weeks, Osman put us on the bus to Istanbul for our flight. I kept in touch with both Osman and Muammar. In 2004, when I got my NSF grant, I asked Osman whether he would like to come back to the United States for three years as a postdoc. Remember, Osman was praying for my grant. I told him that his prayer was answered. Osman was still doing his PhD at Gazi University. He said it will take a year to complete his PhD. It took almost a year to get his H1B visa. I picked the family up from the airport, and they stayed with me for two weeks until they found a suitable apartment housing on the WSU campus. The next day Osman's family arrived, I had a conference. In the meantime, I had taught Sila, their baby. During my absence, she kept asking, "Where is Baby?"

One day, I was invited to eat at their apartment. As soon as knocked on the door and Osman opened the door, Sila said all sorts of things in Turkish, waving her right-hand forefinger. She was outraged. I asked Osman, "Why is

Sila angry?" Osman laughed. He translated why Sila was upset and what she said. She asked her father, "Why Baby did not visit them all these days?" Sila was a beautiful and cute girl who talked a lot. She was attending the WSU early childhood school.

Osman was involved with all aspects of the NSF-TITiC research. He established very close relationships with teachers and students in the schools that participated in our NSF project in Detroit. He co-taught a research class to high school students for the whole semester in Southgate Anderson High School on every Wednesday early morning and helped other teachers and their students after school time such as Grosse Pointe High School. Besides, Osman took night classes on accents, speech, and communication from the English Language Institute at the WSU each Wednesday. Finally, with the approval of the head of the department, he taught an elementary science method course at the WSU. Osman presented a paper at AERA on multiple intelligence and got the best paper award in SIG in 2006. He was also given travel funds. What he did not do was to write any articles from our research, although we had all the data. Instead, he was busy writing and publishing articles on his own; he had completed eight articles. Firat University wanted him back at the end of the second year. I told him not to go because he had an H1B visa to stay at WSU and also look for a position in the United States. In fact, I offered to help him secure a job at a research university. He was stubborn to go back. There were three reasons: Firat University wanted him back. He loved his country, and his family was there. He also had to serve in the army for a year. At the end of the second year, Osman left the United States with his family. One month just before Osman and his wife left, their son, Ali Mert, was born in Henry Ford Hospital in Detroit. So, they went back to their country as a family of four with their newborn son.

As soon as he went back, he applied for his associate degree to the Council of Higher Education. He submitted to the Council a big folder of what he had done with me and the articles he had published. Osman is a convincing talker. He did well at the interview to get his associate professor degree or rank in 2008.

In 2009, I was invited by the 2009 ESERA Conference, Istanbul, Turkey, to give a workshop on the TITiC Project. The ESERA Conference was sponsored by Bahcesehir University, and I was a part of the scientific committee. Osman and I were asked to give a workshop to all science teachers of Bahcesehir Colleges. The Chairman of the Board of Trustees of Bahcesehir University after their fasting gave Osman and me, as well as Norm Lederman and his wife, a big dinner with all pomp and glory. Osman was seated opposite the chairman, Enver Yucel, and they were talking in Turkish. At times, Osman would translate to me the key points. He told Enver Yucel about his postdoctoral research. Enver Yucel, one of the leaders directing the education system in Turkey, did not want him to teach at Firat. He told him, "you are way beyond capacity and find a teaching position at a University in Istanbul." He took us in his own boat to a small palace along the Bosphorus to have dessert.

After the conference was over, I visited with Osman and his family at Firat University, Elazig, Turkey. I gave a University-wide lecture on the TITiC Project, which was picked up by the local TV. I gave a graduate seminar on the Common Knowledge Construction Model (CKCM). I met almost all of Osman's relatives during that time and participated in their festivals. I stayed at Firat University for three weeks to write at least three papers, and we published the first two papers. Osman wanted to take the lead on the third paper. But it never happened. So I took the lead and published it more recently. I will refer to these three articles in a subsequent chapter.

Osman was appointed as the chair of the K-8 science curriculum committee by the Ministry of National Education in 2012. Osman left Firat for Usak University so that he could be close to his parents. The administration made him the vice-rector. He stepped down after two years because he did not want to stay away from active research.

We are still connected. We help each other with articles. I seek help from Osman with my students and my quantitative component mixed-methods research. The last time I saw Osman and his family was when I went to give a keynote address on quality assurance in Istanbul in 2017. I did talk about the work we did and continue to do. I told the audience, that is, quality assurance. You don't wait for the ministry or council to dictate to you. Rather quality assurance should be a way for academic life! Osman is a full professor. Alas, he regrets not listening to me to stay in the United States. He states that it is not challenging to do science education research because no one wants to do collaborative work, and he is by himself. Osman's wife completed her PhD upon her return to Turkey. She is now teaching at the same University.

3.11.2 Muammer Çalik

I made it a point to visit Muammar and his family at Karadeniz Technical University when I was at Elazig. I took the night bus from Elazig and reached Trabzon in the morning. I was planning to stay with Muammer's family for two days. Muammer's wife, Zerrin, told me to extend my visit because I had some spare time prior to my Izmir visit. I knew that I will have to pay extra money to rebook the ticket to Izmir.

I inquired whether there is a granting agency. Muammer talked about TUBITAK. I said, "Let us write a grant proposal to research on elementary preservice teachers using technology." In that one week, Muammer arranged for me to give a seminar on Science Education to faculty staff at Karadeniz Technical University, Trabzon, Turkey.

TUBITAK funded $150,000 to Muammer's 18-month project: Engaging preservice teachers in environmental chemistry learning with innovative technologies and I was the consultant. TUBITAK supported my trip to work with Muammer in 2011 and twice in 2012. Based on this project, we published a book chapter in 2018 as equal authors: *Innovative Technologies-Embedded Scientific Inquiry Practices: Socially Situated Cognition Theory*. Robert Zheng, Editor *Strategies for Deep learning with Digital Technology: Theories and Practices in Education*. Nova Science Publishers, Inc. Included is the model I created for understanding socially situated cognition based on a literature review (see Figure 3.1). This model was influenced by Jean Lave and Etienne Wenger's *Situated Learning: Legitimate Peripheral Participation* that calls for "third space," and situated social cognition by Eliot R. Smith and Gün R. Semin, *Advances in Social Psychology*.

We also published two articles in the *Journal of Science Education and Technology*. One was on improving science student teachers' self-perceptions of fluency with innovative technologies and scientific inquiry abilities. The other was on effects of environmental chemistry' elective course via technology-embedded scientific inquiry model.

Besides TUBITAK work, we published *A review of solution chemistry studies: Insights into students' conceptions*. We also published *Analogical*

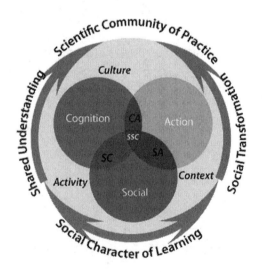

FIGURE 3.1   Socially situated cognition. (From Çalık, M., & Ebenezer, J. (Equal authors, 2018). Innovative technologies-embedded scientific inquiry practices: socially situated cognition theory. In R. Zheng (Ed.), *Strategies for Deep Learning with Digital Technology: Theories and Practices in Education*. New York: Nova Science Publishers, Inc.)

FIGURE 3.2   St. Peter's church – the first Christian Church founded in Antioch.

reasoning for understanding solution rates: students' conceptual change and chemical explanations. On one of the TUBITAK trips, I went with Muammer to a local science education conference. I could not understand the presentations. But it was great to be with Turkish scholars. Most of the time, I was editing or language polishing their papers for publication.

As we know, Turkey is the cradle for Christianity. I visited St. Peter's church in Antioch. This is where the first Christian Church was established. It was a cave. There was a concrete chair on which Apostle Peter was supposed to have used to talk to the group. One of the scholars who was with me took a photo of me in that chair and remarked humorously, "The Queen of Science Education" (see Figure 3.2).

3.12 FIRST SABBATICAL

As part of my first sabbatical in 1998, I spent six months at the University of Bangladesh and Seminary (AUSB). Dr. W.R. Dennis, President of AUSB, A White American missionary invited me as a visiting scholar from May to

December. During my sabbatical, I conducted science and technology education seminars. Elementary preservice teachers took the methods course on science and health with me. *The Co-construction of Science Teaching Conference* was organized by me. I was the founder and fundraiser for the *Children's Science and Technology Learning Center* at AUSB. My mother sent the first $100. In collaboration with Litton Halder, associate professor at the time, I conducted research in chemistry education. In December, I was granted the honor of giving the baccalaureate address for the 1998 convocation ceremony.

My husband was already at AUSB as a missionary. I wanted to be there to help the university. We were given the missionary quarters.

Our house was opposite the president's house. I enjoyed living on the campus amidst the faculty, students, and local people. Concerning the research group I started, we came together each week to talk about our experience. One week, the president's wife shared what happened in her grade four class. She had told the children that they speak with an accent. Immediately, one boy raised his hand, got up, and told her, madam, we don't have an accent. You have an accent because we all speak the same and yours is different. We were all amazed how a fourth-grade student was so smart to give a sensible answer.

I also had a small singing group because I took four years of voice. I passed on some knowledge to this group of vocalists who often performed in church. During my stay at AUSB, I was also their church pianist.

As stated above, I taught a small group of elementary preservice teachers in a grade three classroom how to teach science and health. They really enjoyed that class because the course was conducted in a grade three classroom. I modeled a unit on electricity from a constructivist approach. It was interesting when we learned the structure of a battery cell. With the elementary preservice teachers observing, I asked a grade three children to look at a battery cell and tell one thing about it. Each child gave a piece of information. We put the students' ideas together in a paragraph. The preservice teachers observed how much the children knew. They also witnessed how I was putting their thoughts together in their words.

I had another group of in-service teachers meeting each week. I took them through contemporary methods of teaching through children's ideas, stories, poems, etc. One teacher stated, "So we have been teaching science all wrong."

It was so pleasant to work at AUSB. I would get up early morning and go to the campus to be with the students. I actively took part in most of their events. The teachers visited us often in our home and felt comfortable to walk in. However, a missionary house is not privy to the locals. I served vegetarian food, and we had a lot of fun together.

The president made arrangements to meet with educators at the University of Bangladesh so that I can give a talk. Dr. MD. Azhar Ali, Institute of Education and Research, Dhaka University, Dhaka, invited me to conduct a seminar on research in science education. This group of educators wanted me to give them more seminars. Still, I could not accept their invitation because of time.

My family put several AUSB students up to the master's level in a private school at Spicer Adventist University in India. On a Saturday, my husband and I took the bus to visit a park. As we were strolling, I noticed two young women picking wildflowers. I was wondering why. They approached me after a while and said, "This is for you. We recognize that you are not from here, and we want to show our friendship to you." I was pleasantly surprised. Another day, my husband and I got into a crowded bus, and we had only a standing position. A woman held me so tight. When she got down, she motioned to another bus rider to hold me. We still keep in touch with our Bangladesh friends through Facebook and email.

I took a quick trip to India from Bangladesh by ship and train. I went with my niece, Vinolia, to Bharathiar University, Coimbatore, Tamil Nadu, India, to observe the kind of work she was doing in her Zoology PhD program. I visited with Dr. R. Ananthasayanam (Professor and Head of the Department of Educational Technology), Bharathiar University, to discuss his work. That day Dr. Murgan, Vinolia's PhD advisor requested me to give a lecture about US science education to graduate students and professors in the Zoology Department. During the same trip, my husband and I visited Lady Doak College, Madurai. Immediately, Dr. Nirmala Jeyaraj (Principal), Mrs. R. Victoria (Head of the Chemistry Department), and Mrs. O. Rawlin (Head of the Zoology Department) arranged for me to give a lecture on research in science to more than 100 undergraduate science students. We stopped at my best friend's place in Chennai. Daisy Dharmaraj MD, Director of PREPARE, India Rural Reconstruction and Disaster Response Service, Chennai, Tamil Nadu, India. After a talk in the school assembly, I conducted an all-day seminar for teachers about contemporary teaching and learning methodologies in science.

3.13 INTRO TO ARGUMENTATION

From Bangladesh, I went to Nashville for the rest of the sabbatical. Dr. Richard Duschl picked me up from the airport and on the way, told me, "Your travel is a sign of becoming a scholar." I was a visiting scholar at Peabody College, Vanderbilt University, Tennessee, from January to May, 1999. I gave a talk on *Phenomenography: A research tool for learning how to teach teachers* as part of their Brown Bag Seminar Series. I participated in the implementation of a SEPIA unit on flotation in a grade seven class using *Knowledge Forum*, a computer program. I also took part in argumentation research. It was Richard Duschl who introduced me to argumentation, a pedagogy that was becoming very popular in science education in the Western world. Science education researchers followed Toulmin, a British philosopher. Many conducted research on Toulmin's Argument Pattern (TAP), noted in his book *Human Understanding*.

Researchers published both theoretical and empirical articles in US-based premier journals such as *Science Education* and the *h*. Roselyn Driver and Jonathan Osborne, English science educators, and Richard Duschl from the United States were the first authors on argumentation. Critiquing TAP that it did not satisfy the oral discourse, Duschl asked me to read a book entitled *Presumptive Reasoning* by Douglas Walton. Figure 3.3 is a combination of Toulmin's argumentation pattern and Walton's presumptive reasoning.

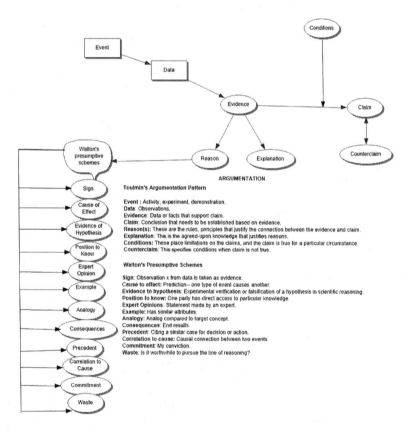

FIGURE 3.3 Toulmin's argumentation pattern linked with Walton's presumptive reasoning.

On the first page of the book, I noted that the author on the schemes of presumptive reasoning was from the University of Winnipeg, Winnipeg, Manitoba. I was so excited because he was my neighbor. I had to be introduced to Walton's work at Peabody Teachers College. When I went back to Winnipeg after my sabbatical at Vanderbilt and Maryland University, I visited Douglas Walton. He spent a lot of time with me. He stated that his work on argumentation is in law. I told him that I am going to translate his work on presumptive reasoning to science education. He was not sure how that was possible. Together with my graduate student, I published an article using Douglas Walton's framework on presumptive reasoning. Walton's presumptive schemes useful to science education are as follows: Sign (Observation X is taken as evidence); Cause to effect (Prediction—one type of event cause another); Evidence to hypothesis (Experimental verification or falsification of a hypothesis in scientific reasoning); Position to know (One party has direct access to a particular knowledge); Expert opinion (Statement made by an expert); Example (Has similar attributes); Analogy (Analog compared to target concept); Consequence (End result); Precedent (Citing a similar case for decision or action); Correlation to cause (Causal connection between to events); Commitment (My conviction); Waste (It is worthwhile to pursue the line of reasoning?).

One day, Richard Duschl asked me during our conversation, what is my academic goal? I did not understand what he meant by that question. Then he said whether I am aspiring to become a leader at NARST or AERA, the two associations for science education. I told him, I have no ambition of that kind. My main goal was to nurture our only child. All these activities will distract me from raising our promising son. However, I was first voted as the program chair. The following year I was elected as the chair of the AERA SIG – Science Teaching and Learning. The past chair asked me how I was able to increase membership from a handful to a few hundred members. The secret is to allow everyone – you reach. Since 2009, I have not attended AERA or NARST. But I do get emails asking me to join again.

There are several reasons for not attending these conferences. I was more interested in traveling the world, which I have done and reached out to science education scholars. In fact, when I gave a talk at the NSF about the Common Knowledge Construction Model, one officer asked me how do I travel the world? Another issue with AERA and NARST is that I have found some members talking down to me. The same people value me only after they see me again at Science Education and or JRST Receptions invited by journal editors. University funds stretch only for attending one conference. Since 2009, I have felt I should be closer to my family of trinity – father, mother, and son. Although I don't attend these conferences, I encourage all my graduate students to become AERA and NARST members and attend meetings.

3.14 WORLD BANK MISSION

It was my desire to go to the World Bank and request funds for my research projects. I heard that a woman from the World Bank had come to Peabody College, Vanderbilt University, to recruit individuals to go on World Bank education missions. I was excited. I asked the secretary about the details. The secretary told me that she was here and she had gone back to Washington, DC. I asked the secretary for the email address and the phone number. She was rude and did not want to give me the information even after telling her that I am a visiting professor. Finally, I was able to get the information I needed.

I sent a message to the World Bank woman stating that I would like to visit her. She was excited first. Then she followed up with another email stating not to expect her to fund my travel to the World Bank. After all, my sister and many cousins lived in Maryland. I could stay in their homes. Moreover, I am not a stranger to DC. I wrote to her to say that I was coming there anyway and I will drop in at the World Bank. She gave me an appointment. I took the two books and articles that I had published to show her. The lady was very kind and was impressed with my work. I would have chatted with her for more than an hour. I came out of her office. One of the administrative assistants, Energy James, asked me for my CV and gave it to the main boss, who was next door. He told his secretary to take me to see every officer that represented the various countries. I stopped at the first office. We chatted for some time. She asked me whether I would give a talk at the World Bank. Using her computer, I wrote about the topic of my seminar. She sent it to the education group from various countries. Then she told me not to expect too many individuals at the meeting because of the last-minute announcement. I was giving the talk the following morning at 10 a.m.

I spent the night with my first cousin Merlin Ponraj at her home because she worked at the World Bank in another division. I told her about my experience at the World Bank. Then we talked for a long time. She asked me whether I was ready with my talk because she did not see me preparing. She said it is a serious business. On the following morning, I wrote a page to distribute as a handout.

My cousin took me and left me at the office I was supposed to be. The officer and I walked into the room of more than 30 people. Another first cousin who worked at the World Bank came for the talk, but Merlin did not attend because she said we look alike and it was best for her not to come.

The officer introduced me. I gave a 45-minute talk about students' conceptions of the salt-water system and the implications for teaching and learning. After answering many questions from the much-interested audience, the officer and I had lunch together. She told me that she knew it would be an "excellent" talk, but she did not expect to be "outstanding."

I went back to Peabody College. On the next day, Richard Duschl asked me about my experience at the World Bank. I told him nothing much because I went there to get research funds. After a few days, I received an email from the Indonesian officer of the World Bank requesting me to go on a three-week mission with her. I was ever so excited about this opportunity. Who was the first person to tell? Richard Duschl.

Rick arranged for me to go to the University of Maryland and spend a month with David Hammer, a physics educator. I went to the department and David was not there. There was another professor present. I told him about the purpose of my visit. The first thing he did was to check me on Google Scholars. After that, he was ready to talk with me. It seemed to me people were selective about who they would speak and even spend time with. While I was with Dr. D. Hammer as a visiting scholar in June 1999, I had the opportunity to conduct a seminar on phenomenographic categories for common knowledge construction. I used chemistry as an example.

I had an email from the World Band about my salary, and I asked the secretaries how much their professors ask. They counseled me to ask for $800/day. I felt a little awkward to ask. But the officer from the bank told me that they will pay me $500 with no tax deductions. I would have gone without a fee. This was all new to me. After a quick negotiation, the officer who contacted me asked me to go to a particular office in the World Bank to get my first-class air ticket, and the $4,000 for day-to-day spending. That week I left for Indonesia. It was my first first-class plane ride, and it was excellent.

The World Bank allowed me two stopovers, one going and one coming back. I stopped for a week in the Philippines. I visited the University of the Philippines in Manila. The science education colleagues embraced me because they told me that they know me through the literature. I was impressed with their science education lab facilities that South Korea had built for them. I visited a few places of interest. I spent a whole week at a Christian university. I was asked to stay in a White American Missionary home. The couple was old. The woman asked me to sleep on the floor in a tiny room with shelves around the walls among a pile of scattered books. The next morning, she asked me to edit a dissertation because she was the editor of the PhD students' writings. Besides editing, I made comments about the content and format, which she did not appreciate.

She made some nasty remarks. She also told me in passing that because you are an Indian, you slept on the floor. A white person would not have slept on the floor. I met a Canadian white couple on campus. They asked me to stay in their home. It was peaceful and we had great conversations. I walked every morning around the school. No one talked. On the last day, another white

Canadian professor who I knew in Canada was teaching in the Education Department. He asked me whether I wanted to give a lecture in his class for an hour. He said that the topics are on Piaget. He told me strictly and firmly that he will provide me with only an hour. I taught his class.

Yes, I talked about Piaget and told the class members that one of my articles is archived in Switzerland in Piaget's library. At the end of class, the professor asked how long I am going to be there, and he said my discussion was thought-provoking and wanted me back to teach his students. I told him I was there for the whole week. No one recognized me except for one African graduate student who asked me about my nationality, I told him am a Canadian. He said I did not look like one. He said, "The color of your skin tells me you are not a Canadian." Then I gave him a piece of my mind. That PhD student was not convinced.

I flew to Jakarta. The World Bank officer came to pick me up in her BMW car. While going to the Four Seasons Hotel, she told me that she was expecting a white woman with gray hair. She also asked me how old I was because she said I looked very young. A lady in uniform took me to the floor, one below the highest in the hotel. On the top floor, there were free snacks and soft and hot drinks that we can help ourselves. She also showed me the spa and fitness facility. I was wondering why she was with me for more than half a day. She told me that she wants to make sure that I am settled comfortably.

The hotel room was huge, with marble floors and a basket of fruit. In the hallway too there were baskets of apples, oranges, and bananas. While staying there for two weeks, I would take the fruits that were placed in my room and give it to beggars who were in the vicinity of the Four Seasons Hotel.

The World Bank officer gave a big pile of reports that described the $56M project called PQUIP. She said that I had to do their final evaluation of the seven-year program and submit a five-page report. During the two-week, she and her colleagues drove me to visit the Department of Education, their school examination central office, many primary schools, principals, and superintendents. I conducted many interviews, transcribed them, and wrote a report based on the data that I had gathered. Two women took me to Sumatra for a whole week, where I conducted the same activities.

Before I went to Indonesia, I was wondering how they will treat me as a South Asian from Canada. Will they treat me just like they would treat a white person who goes from North America? The local officers took a lot of pictures and talked to me cordially. In the last school I visited, the principal wanted to take a picture of me with her teachers and staff. As parting words, she told me, "You are one of us." These words pierced my heart and soul and still remain with me.

I had an encounter with the fellow who came as a consultant from England to Indonesia for several years. He told me, the World Bank is stupid to bring people who don't know the language. I knew he was attacking me. The best thing is not to retaliate in such circumstances. On my return trip to Canada, I stopped in Singapore for the weekend. I visited their zoological gardens and many other places of interest. I also crossed the border and went to Johor, a neighboring city in Malaysia. I flew to Hong Kong, and in the transfer at the airport, I left one of my pieces of luggage on the top of a steep escalator. There is no way of going up to get it. So I walked and walked until I reached the airline counter. I waited for a long time to speak with an officer. Finally, I talked to a young woman. She was rough and rude. She walked with me in search of my luggage. We walked and walked. She was miserable and would not talk to me. She was angry, and I could tell that from her behavior. One time she stopped and asked me how did you cross security. Of course, no one stopped me. Then she was surprised that I had walked a long distance. I told her I walk a lot; this is nothing. I am simply worried about my luggage.

We found the luggage. It was still where I had left it. Then the airline agent took me back to the Cathay Pacific Airlines counter by train. I was also surprised that I had walked a long distance. When we reached the desk, it was boarding time. She made me stand in the ordinary lane and then asked me for my ticket. When I gave her the air ticket, she was shocked. She gave me a smile that was not there while searching for my luggage. She said, "Why didn't you tell me this?" I was watching her disposition and said, "Father, forgive her for she knows not what she is doing to the ordinary passengers."

The following year, the Indonesian World Bank requested me to come to teach teachers. I was happy to go, but she sent me a note to say that I was too expensive and they want local experts to take the lead. Through my visits to the World Bank, I have learned that it does not fund research activities. Still, they promote using research-based teaching and learning methods in educational endeavors. I was also a World Bank consultant for several years.

3.15 A ROLE MODEL FOR THE NEXT GENERATION

While I was at the University of Manitoba, a mother in the church told me that she has collected all my pictures from the local newspapers. She showed the album to me. I was surprised. I asked what motivates her to do this. She told me that she wants to show her daughter how women can achieve anything they want, and I was one example.

When I left Manitoba, there was a big farewell party. People gave testimonies. In my remarks, I told them that I was not active in the church. Still, they were gracious to provide me with a friendly farewell, and I appreciate their generosity. One Filipino-Canadian got up and said that I hired their children to do academic work. Also, they said that I was a role model for their children, and that was a sufficient contribution to the local church.

The Faculty of Education at the University of Manitoba also gave me a farewell. The head of the department, Professor Dexter Harvey, who hired me, gave the farewell speech. He made kind comments. One comment still remains with me. He said, "One thing I observed was that over the last ten years she stayed with us, I have never heard her gossip." But there was a woman who told me, "So you are going to your own people." Another remarked, "You will have to wear a bulletproof vest." For one moment, I have not regretted the move to Wayne State University. Here I am 20 years later!

The Science Banqueting Table

4

4.1 FACETS OF CONCEPTUAL CHANGE THEORY

Based on my extensive reading of scholarly literature during my doctoral study, I accepted conceptual change as a viable theory for improving science teaching and learning. Scientific knowledge growth underpins the first conceptual change model. This conceptual change model was advanced by Posner, Strike, Hewson, and Gertzog (1982). These authors postulated four general conditions for conceptual change: "dissatisfaction, intelligible, plausible, and fruitful." For example, a student is dissatisfied with the initial idea: A person can see an object because the eye sends out streaming particles to it. Students learn through inquiry that the rays of light travel from the object to the eye, which becomes intelligible. The learned idea becomes plausible based on several tasks that show the latter to be true. When the student realizes that the newly learned idea can be used across different contexts, the idea becomes fruitful.

A philosopher, Paul Thagard (1992), viewed the condition of replacement and/or abandonment as a criterion of conceptual change. Thagard used a fascinating metaphor to explain his theory of conceptual evolution. If a person moves from one branch to another in the same tree, one idea is replaced with another, remaining in the same camp. A person moving from one branch of a tree to another tree means he abandoned the original idea and moved from the first camp to a different one. Educators, Chi and Roscoe (2002), redefined the notion of conceptual change. They considered only the process of repairing misconceptions and the ongoing development of preconceptions as conceptual reorganization, revision, or accommodation. They believed that altering a naïve conception or misconception is the role of conceptual change learning.

In contrast to the above models, in line with Marton and Booth's (1997), Ivarsson, Schoultz, and Saljo (2002) argued that naïve conceptions do not serve a purpose in conceptual change. These authors situate this position in a modelling of conceptual change as the appropriation of intellectual tools. According to the Swedish phenomenographic perspective, variation in the critical aspects of the object of learning needs to be experience for a constitution of learning to take place (this is often referred to as the Variation Theory of Learning — for a comprehensive overview of this model of learning and empirical evidence for its educational effectiveness, see Marton, 2014 and Marton & Booth, 1997). Using this phenomenographic perspective several well-argued cases have been made that such learning is always contextually situated (for epistemic examples, see Linder, 1993 and Linder & Marshall, 2003). The starting point for these cases is that, from a phenomenographic perspective, what is learnt has both structure (a *how* at tribute) and meaning (a *what* attribute) and these attributes are embedded in situated person-world relationships. In 1995, building primarily on Linder (1993) Jim Gaskell and I pushed the idea forward that experiences of contextually stimulated variations resulted in what we characterized as "relational conceptual change."

Once when I was explaining to a group of Christian professors in a Christian College, Nirmala, who did not have sight, a professor of Biblical studies told the audience, conceptual change is not only evolutionary and revolutionary, but also "revelationary." This took me by surprise and challenged me intellectually. I have not still probed this conception of knowledge growth, but it is a wonder.

All conceptual change models follow a set of core tenets. Every student has the opportunity to explore his/her ideas and express them from their own viewpoint. Teachers become aware of students' ideas, and students become aware of their ideas. Teachers integrate students' ideas explicitly and systematically into the curriculum. Students acknowledge their ideas and expose them to peer scrutiny.

When students' conceptions are the basis for curriculum, teaching, and learning, it is apparent that we value their voices. Teachers engage students in inquiry activities to negotiate scientific ideas. Evidence and explanation underpin students' developing ideas. Students develop reasoning and communicative skills as they articulate their thoughts. They also have the opportunity to monitor their morphing ideas.

4.2 VARIATION THEORY OF LEARNING FOR RELATIONAL CONCEPTUAL CHANGE

Marton (1981) introduced a more radical view of conceptual change. He assumed that knowledge originates in the person–world or subject–object duo. Rational relationships give rise to variations of thoughts. He coined the *Variation Theory of Learning* as phenomenography.

Phenomenography is an experiential perspective of conceptual change. Conceptions of phenomena are relational, that is, phenomenography describes the relation between the individual and the phenomenon. Phenomenography shows concern for both the subject (how) and the object (what). It adopts the principle that human thinking is contextually determined. The assumption is that conceptions of reality do not reside within individuals (intellectual capacity or developmental stages), as Piaget believed. In contrast, phenomenographers view people's notions of truth as particular-to-particular context and problems raised within that context.

According to Piaget, people's understanding of the world differs because of their age. It either depends on the developmental level, or is connected to variations in the intellectual maturity of the learner responding. This implies that there is a systematic quality to students' answers according to their maturity level. The phenomenographic approach is more concerned with describing the possible variations in conceptions individuals express for a particular phenomenon. There is no strong concern for the developmental mechanism that created that variability. Phenomenography adopts a worldview that the world is inherently multifaceted and open to variations in interpretation.

Phenomenographers believe the world is seen through a particular "lens." That there is no such thing as common and reality to every human. Marton argues that individuals believe that the most original images of our world are always given. And the imaginings are mostly present in individual consciousness. Instead, according to Marton, how we view are reflected in the way we organize society. Marton advocates that we must look beyond the individual in our search for understanding the various ways in which people perceive a phenomenon. In this view, conceptual change results from changes in the way students use intellectual tools in multiple contexts. These principles set apart phenomenography as a distinct research specialization.

Phenomenography is recognized as complementary to conceptual change theories because of Marton and his group's view of learning as a change from one way of understanding a phenomenon to another and qualitatively different way of understanding the same phenomena. This statement implies that students need to undergo conceptual dispersion and should be able to distinguish between provinces or contexts of meaning (Linder, 1993; Linder & Marshall, 2003). For example, we need to help students talk about differential air pressure instead of the vacuum cleaner sucking air. Dissolving is expressed in ionic theory as opposed to melting. Shadow formation is blocking light, instead of emitted from the body. It is "okay" for learners not to abandon their everyday talk: Vacuum cleaner is sucking up dirt. Salt is becoming a liquid. Shadows are a part of me, light ejects it and carries it onto the screen. A learner uses concepts specific to the context. Conceptual frameworks are not abandoned, instead they are used appropriately within context. Learners are taught to distinguish between the disciplinary background and everyday context. Such theoretical considerations support the contention that conceptual change could be successful with students studying science. I now turn to the discussion of a teaching model that I developed from phenomenography that advocates relational conceptual change.

4.3 COMMON KNOWLEDGE CONSTRUCTION MODEL OF TEACHING AND LEARNING

The Common Knowledge Construction Model (CKCM) serves as a model for teaching and learning (see Figure 4.1). It is rooted in phenomenography, the *Variation Theory of Learning*. Common knowledge in science means constructing a reality that resides in the science context of meaning. This reality differs from those employed in everyday thinking or thinking in other settings.

The CKCM consists of four interactive phases of teaching and learning: Exploring and Categorizing; Constructing and Negotiating; Translating and Extending; and Reflecting and Assessing. Phenomenography, a branch of conceptual change research specialization, guides all four phases of the CKCM.

Conceptual change theory, rooted in Piaget's work, advocates that knowledge resides in mind. Whereas, phenomenography espouses the relationship between the subject (mind) and the object (e.g., the concept of excretion). Although the theoretical base of the CKCM is phenomenography, the model borrows learning strategies and tools from researchers' work rooted in the

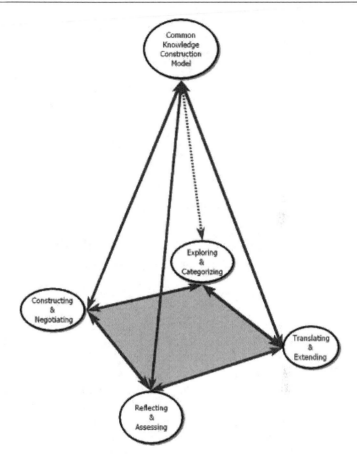

FIGURE 4.1   The Common Knowledge Construction Model (Ebenezer & Connor, 1998; Ebenezer & Haggerty, 1999).

conceptual theory of Piaget. The CKCM adopts the learning outcome, that is, move toward scientific explanation and practices. This process is rooted in conceptual change theory. Therefore, I argue that CKCM is situated at the intersection of two approaches. The conceptual change theory rooted in Piaget and phenomenography anchored in Marton.

The first phase of the CKCM is *exploring and categorizing*. It uses phenomenography as an inquiry tool to generate students' conceptions of natural phenomenon or science concepts. Exploring students' ideas using one or two related simple tasks is functional. This process is sensible from the learners' point of view. Students' multiple meanings are encouraged. Students' expressed thoughts are not judged for being right or wrong, as would occur in a diagnostic or deficit model. Just as scientists explore their various ideas, students are provided with opportunities to explore and become conscious of their personal views. In doing so, students begin to understand that science is an attempt to explore and explain natural phenomena.

Found in the pool of students' expressions are personal ideas with inter and intravariations. Intervariation denotes different ways of thinking between individuals. Intravariation means the different meanings within an individual. We identify commonalities in inter and intravariations of meanings and develop into "phenomenographic categories." The categories of description are ways of denoting the researcher's interpretations of students' conceptions of a particular phenomenon. Categories of description feature qualitative and quantitative aspects. The qualitative outcome is the categories of description. The specific quantitative result is the frequency distribution related to the categories. To construct common knowledge in science, the categories of description provide a basis for lesson planning and classroom discourse. Personal ideas are shared in class so that peers can evaluate the merits of these ideas in an open forum through a process of construction and negotiation.

The second phase, *constructing and negotiating*, directs students to specific ways of construing reality. This phase develops the theoretical accounts of nature and society. Construction of scientific knowledge and negotiation of meanings of concepts should occur first before we assign concept labels. For constructing and negotiating to take place between teacher-to-student(s) and peer-to-peer discourse, the teacher acting as a mediator is a procedure. Students make observations using their prior ideas, record relevant thoughts in their notebooks, propose and interpret multiple meanings, and think about their thinking.

The construction and validation of contextual knowledge occur when the teacher and students become collaborative meaning-makers, searchers, sharers, and negotiators. These elements contribute to an attitude of collaborative scientific inquiry. This process of meaning-making suggests that scientific knowledge does not entirely rely on observation, experimental evidence, rational arguments, or skepticism. Instead, it portrays the tentative and negotiating character (social objectivity) of science. Students recognize that conceptual change occurs when they question their original conception based on everyday context. Then they subject their ideas to critical thinking, inquiry, and peer review to be useful for the science context of meaning-making. Students also realize that it takes collaborative time and effort as well as patience and empathy toward fellow learners when formulating scientific ideas.

The third phase of the CKCM, *extending and translating*, enables students to use their conceptualizations of scientific ideas developed in Phase two to shape problems of socioscientific inquiry. By engaging in the investigation of science-related societal issues, students develop an awareness of the complex interactions among science, technology, society, and the environment through a critical thinking disposition. Such a character consists of identifying complex, open-ended problems, exposing individual's views of the problem for critical inquiry and conceptual change, asking essential questions, pondering causes and consequences, and considering alternative positions. Such an educative process is anchored in the sociocultural paradigm. This orientation encourages the expression of multiple views about an issue, focuses on the process or science of learning within the context, and values that learning changes over time. Domain-specific educational tools, talk, signs, and symbol systems, which have a temporal and geographical bearing, shape human thinking. This process calls for changing the classroom discourse practices from information dissemination to dialogic discourse. This transformation portrays the tentativeness and purposefulness of knowledge creation. This change sees science as a human and social activity.

The fourth phase, *reflecting and assessing*, is integral to exploring and categorizing students' conceptions, constructing, and negotiating shared common knowledge, as well as translating and extending students' understanding of science concepts into the study of personal and societal relevant scientific and socioscientific issues. Traditional assessment methodologies are fill-in-the-blank, multiple-choice, true/false questions, and matching questions. These require student regurgitation of information or the right answer. Therefore, these do not best serve as effective assessment practices for conceptual change inquiry teaching and learning. For conceptual change learning, researchers call for alternative assessment (Barton & Collins, 1993; Collins, 1992; Duschl, 2003; Duschl & Gitomer, 1991; Liu, 2004; Micari, Light, Calkins, & Streitwieser, 2007; Novak, 2002; Sampson & Clark, 2008). In a conceptual change inquiry process, assessments should measure how students explore, expose, revise, or reject their conceptions based on evidence and explanation. Teachers should track the small steps that students take to understand difficult science concepts and conceptually change. This tracking helps determine how effective teaching has been for conceptual change. What concepts need to be further explored becomes evident. How students use the concepts to design, conduct, and evaluate scientific and socioscientific inquiries that have personal and societal relevance is part of the assessment. To measure these processes of learning continuously and reflectively, both teachers and students need to engage in formative assessment tasks. These attributes are necessary for students to consider "*how* they know *what* they know…[while establishing]…standards for knowledge claims communicated in science" (Ruiz-Primo & Furtak, 2007, p. 64).

The CKCM integrated assessment strategies include prediction–explanation–observation–explanation (PEOE) strategy rooted in White and Gunstone (1992). A form of PEOE is practiced by the teacher who was involved in this study. Authentic assessments promote and reveal students' conceptual understanding and enable teachers to make decisions for immediate teaching plans (Ruiz-Primo & Furtak, 2007) for developing "learner appropriate sequence of lessons" (Ebenezer & Haggerty, 1999, p. 406). These assessments are consistent with the National Science Education Standards (NRC, 1996).

The four phases of the CKCM reflect the National Science Education Standards, which emphasize the learning of science content (the what and how of science) and using this knowledge (where and why of science) in problem-solving societal issues. The development of what, where, when, how, and why of science is fundamental to scientific literacy.

4.4 COMMON ATTRIBUTES OF CONCEPTUAL CHANGE TEACHING AND LEARNING

Regardless of theoretical suppositions, as described above, the hallmarks of teaching and learning for conceptual change are the same. Students explore their conceptions of a natural phenomenon. They become consciously aware of their understandings. They share personal notions within a learning community for appraisal. Finally, they test and compare particular conceptions with scientific models and explanations for plausibility. Through a social process, students refine, reconstruct, reconcile, or reject personal conceptions to align with the scientifically sound and agreed-upon understanding. These hallmarks of teaching may be adopted by those who understand conceptual change as undoing misconceptions or altering naïve conceptions, as well as by those who value variations in students' conceptions. With intellectual empathy, the advocates of the *Variation Theory of Learning* use students' conceptions as essential frameworks for the development of learning activities that stress on the variation and application of a concept within the science context.

Conceptual change studies have primarily focused on issues surrounding the changing of students' conceptions to those accepted by the scientific community. This is either confrontational or intellectually empathetic. The application of the conceptual change framework on student science achievement is also significant. This is because science educational policies in many countries are calling for dramatic changes in teaching and learning practices so that each child may attain the optimum achievement in science. Such a vision is vivid and alive, including in Canada, India, and the United States, where I have conducted studies. Studies have proved that the relational conceptual change theory improves achievement. In fact, teachers, who I educate, have told me that when they use the conceptual change inquiry of teaching and learning, students' grades improve. Also, teachers have stated that there are hardly any management problems because students are engaged actively in assessment-embedded in-depth meaning-making inquiry activities. They are also involved in monitoring their grades. However, teachers, who are doubtful and do not fully attempt this approach, state they are falling behind in covering the curriculum. What some teachers do not understand is that the fundamental ideas are revisited in science, and thus they are uncovering the curriculum. For example, an ionic equation based on the solubility of an ionic compound in water generated in learning solution chemistry can be applied to units that involve salts breaking into ions. Acids and bases, electrochemistry, and chemical equilibrium are examples of chemistry units that use the concept of dissociation. It is understanding the type of knowledge – macroscopic, submicroscopic, and symbolic – and transitioning among these spaces with ease is vital. Any wonder, Thagard, in his book on conceptual revolutions, through a network of words, eloquently argues science has had only seven significant revolutions. They are oxygen theory, cell theory, molecular kinetic theory, wave theory, plate tectonic theory, and psychological learning theory.

4.5 CONCEPTUAL CHANGE RESEARCH EN ROUTE FROM PUNE TO TAMIL NADU, INDIA

Using the *Variation Theory of Learning*, my first co-author and I designed a study to understand the issues revolving around student science achievement and relational conceptual change in a unit on excretion at a middle school in a private school in Pune, India. How did this study originate? I happened to visit the university where we conducted the study. The Dean of Education asked me to give a lecture on my model of teaching and learning, the CKCM. After the talk, a zoology professor, Sheela Chacko, asked me whether I would help her adopt the model in her classroom. I responded that, if she is willing to research teaching and learning using the model, I can help her.

That whole year I was guiding her from the United States on how to explore students' ideas and collect data in her grade 12 classroom. However, she was reluctant to research in her class because she was preparing her zoology students for the national examination. She did not trust me. So she found a grade seven teacher who had a master's degree in microbiology. In this class, Sheela explored students' ideas of excretion, categorized, and taught lessons, negotiating with them the difference between excretion and digestion. We spoke many times on the phone. What I am about to describe is a sidelight. One day I was up all night talking to her, and the next day I was supposed to go to Ottawa from Detroit to visit a friend. Nearing Toronto, I fell asleep in the traffic jam along Toronto connectors, and slowly the car moved and crashed into the vehicle ahead. The result was several thousands of dollars on fixing the bumper. Without adequate rest, research causes the unexpected.

I invited Sheela the following year (2004) to travel with me from Pune to South India. She was excited because, in our journey, she could also visit her mother, who lived in Kerala. After I gave a keynote address at the Maharashtra Institute of Technology to 300 scientists on the ultimate meaning of life, Sheela and I took the overnight train. I planned this trip so that we have a long journey together to discuss our research. She had brought all the students' papers and artifacts, as well as her instructional data. I explained how we can write an article for publication. After two days of train travel, we reached Chennai at night.

We stayed with my husband's cousin. Because of Christmas season, we could not get a train or bus ticket to travel further south. We wanted to go first to Madurai, spend one day at Lady Doak College, where I got my chemistry degree. Then we intended to proceed to the city in Kerala, where Sheela's mother lived. Without regard to her concern about traveling only in a luxury bus, I took her to the state bus station just to see whether we can ride the bus. Fortunately, the Thiruvalluvar bus with the board to Madurai was waiting with no passengers. We arrived a long time before its departure. Asking the bus driver, we got into the bus with our two big pieces of baggage and headed to Madurai; the bus reached at 12 midnight. Sheela was afraid, and she was calling her husband every few hours. She is Indian, and I expected her to lead me, but her husband told me that I have to hold his wife's hands and take her to the doorstep of her mother's house.

Keeping the promise, we took the next bus from Madurai after spending a night at Lady Doak College to her hometown in Kerala. I was supposed to leave her, and the following day I was catching a flight to Sri Lanka from Trivandrum, a city in Kerala. Tsunami engulfed Sri Lanka, Chennai, and the coastal areas in South India. Sheela 's mother advised me not to fly to Sri Lanka. The images of tsunami on social media were scary. The tsunami wiped away thousands of people, including the people who regularly went for walks along the Marina beach in Chennai. Instead of going to Sri Lanka as planned, the next day, I took a bus to Nazareth/Prakasapuram, my home town, where I was born. My parents and brother's family from the United States were visiting my hometown. They were dedicating a Science Center in a Christian school in honor of my sister, who passed away due to unfortunate circumstances. After I spent the new year with my family, I took the train back to Chennai, and on the way, Sheela got in. We both transferred to the Mumbai Express train.

With the motion of the train, we discussed the paper we were going to write together. I encouraged Sheela to take the lead in writing the article. However, it was challenging to guide her from the United States. Finally, I wrote the paper, and Sheela gave me the data. We also sought help from Osman Kaya, a Turkish Professor, to help us with statistical analysis. The article was submitted to the *Journal of Research in Science Teaching*. We did several rounds of revisions with the associate editor's advice because the paper was two-pronged, professional development and student achievement. We finally focused on the influence of conceptual change inquiry on science achievement. In the end, the associate editor told me, our article was the only phenomenography that the journal had accepted. The editor also stated that the rate of journal acceptance that year was 13%. She thanked me for an outstanding article because phenomenography was explained well, distinguished from other conceptual change models. If not for that associate editor who was kind to help me directly with this complicated article, publishing would have been nearly impossible. This is the second time an editor had helped me. I suppose editors see value in the message I am trying to convey. I negotiated with the editor why our article needs to be published, and I followed her instruction and made the major and minor changes the reviewers had noted. The point here is, I am happy even when my papers are accepted with significant revisions. Then, there is still hope when major and minor corrections are taken seriously.

4.6 SHEELA'S ATTEMPT AT CONCEPTUAL CHANGE RESEARCH ON EXCRETION

4.6.1 Prediction–Explanation–Observation–Explanation (PEOE) Sequence

As an exploration activity, Sheela showed a sandwich and a glass of water. She reminded them about their breakfast, and asked them the following second-order questions:

1. What waste products do you think are produced in the body?
2. How do you think wastes are produced in the body?
   Draw and write how waste is *produced* and removed?
3. Which organs do you think remove the wastes?

In line with this indirect approach, the exploration task and questions should relate to learners' experience (simple and sensory) such as the ones used in this study (i.e., eating sandwich and drinking water).

Students predicted and drew their diagrams individually. Then, they shared their written explanations and diagrams to their peers.

The teacher collected all students' papers. She grouped students' ideas of excretion into phenomenographic categories of description and not the students. She also made an idea splash chart, randomly pasting their prior teaching conceptions.

The next day, Sheela engaged seven groups of students to group their ideas on a chart paper for posting it on the wall. Then, she using the PEOE strategy showing the entire pictorial chart for observation and explanation of the digestion process (see Figure 4.2).

The teacher discussed metabolism and concept of cell. Students made brochures by referring to materials and resources on digestion and excretion. They included a flowchart on what is happening to the food taken in, how waste is produced, and make a diagram of excretion. Students formally presented brochures to the rest of the class.

The teacher's exploration and subsequent instruction reveals how students' conceptions can be logically connected to a more complex phenomenon such as the cellular processes of excretion.

We charted the categorization made from students' pre and postteaching ideas and matched to observe conceptual change. Students' answers were expressed in one word such as "urine," in a series (e.g., several types of wastes – urine, sweat, feces), words strung together in a proposition (e.g., the waste products produced are urine and toilet), and several statements. We placed the answers to question two into three broad categories representing a hierarchy of knowledge, from simple (sensory) to complex (cellular) (see Table 4.1). Once we developed the categories of description, we counted and noted the frequency distribution between the pre and the post.

4.6.2 Conceptual Changes in the Study of Excretion

Components arising from qualitative data include: (1) the addition and deletion of categories of ideas from pre to postteaching; (2) the change in the number of students within the categories of ideas; (3) the replacement of everyday language with scientific labels; and (4) the difference in the complexity of students' responses from pre to postteaching.

The Addition and Deletion of Categories of Ideas from Pre to Postteaching. Table 4.1 section a represents six categories for the types of waste products (CQ1). Two categories – nitrogenous waste and metabolic waste – are generated only from postteaching conceptions. Table 4.1 section b indicates that the postteaching ideas include the addition of cellular process to describe how waste is produced (CQ2). Skin and lungs are added as organs producing waste, but the digestive system is abandoned (see Table 4.1 section c – CQ3). It is obvious that changes have occurred from pre to postteaching, which can be used to infer the nature of the conceptual change. For example, the addition and deletion of categories reveal the change in students' understanding because prior to teaching students used everyday language such as "urine" and "feces" (see Table 4.1 section a). After teaching, some students replaced their everyday language with equivalent scientific terminologies such as "nitrogenous waste and metabolic waste." Common word replacement with scientific language points to the notion that teachers should first explore the language that students commonly use before teaching the scientific language. I believe that when this initial step is taken students will develop a better understanding of new concepts.

The students know that waste products occur as a result of eating. They have also learned the functions of digestive and excretory organs. More importantly, students are aware that waste is processed at the cellular level (see Table 4.1 section b). Students are also aware of skin and lungs as organs that produce waste (see Table 4.1 section c).

The Change in the Number of Students within Categories of Ideas. As represented in Table 4.1 section a for CQ1, there is a negligible difference between pre and postteaching ideas for both urine (16, 18) and feces (19, 21).

TABLE 4.1   Descriptive categories of 33 seventh grade students' pre and postteaching conceptions of excretion

	PRE	POST
(a) Waste products		
Urine	16	18
Feces	19	21
Gases	1	5
Sweat	1	16
Nitrogenous waste	0	5
Metabolic waste	0	3
(b) How waste is produced		
The eating of food and digestion	22	22
The functions of the kidneys and digestive organs	4	9
The cellular process	0	4
(c) Organs removing waste		
Anus	10	23
Kidney	6	21
Intestines	11	3
Digestive system	10	0
Skin	0	18
Lungs	0	3
Urinary system	10	2

This is not surprising because students biologically experience the phenomena of excretion and digestion. Prior to teaching, only one student considered the phenomenon of gas production as waste. However, in the postteaching phase, five students mentioned specific gases such as carbon dioxide and ammonia as excretory wastes (see Table 4.1 section a). This is because the experimental teacher explicitly drew students' attention to the excretion of carbon dioxide through a PEOE lesson (see Figure 4.3). The experimental teacher's explicit teaching of the biological ideas contrasts with Din-Yan's (1998) statement, "When explaining the mechanism of gaseous exchange, teachers and textbooks seldom state explicitly that carbon dioxide is a metabolic waste or refer to the process of removal of carbon dioxide during breathing as excretion" (p. 109).

Sweat is another waste product students listed for CQ1. Prior to teaching, only one student thought of "sweat" as a waste product. However, after teaching, 16 students considered "sweat" a waste product. Children usually overlook sweat as waste because its production takes place automatically, but urine

(a)   (b)

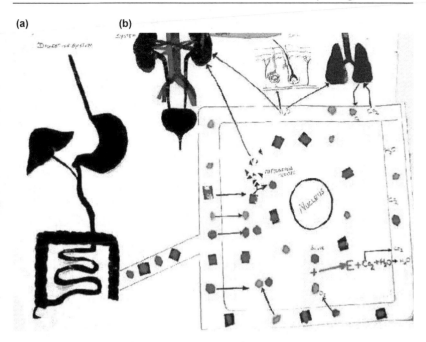

FIGURE 4.2 A diagram for the PEOE strategy to help students (a) predict and (b) explain what might happen to the digested food.

and feces are readily accepted as waste products because discharge of these is readily visible. Because the experimental teacher brought learner awareness through a process of negotiation with the PEOE strategy (see Table 4.1), there was a dramatic increase in the number of students considering sweat a waste product. For instance, students became aware of the reactions taking place with the absorbed food at the cellular level. They also learned that blood transports the metabolic waste produced in the cell to the kidneys as urine and to the skin as sweat, which consists of water and salts.

Table 4.1 section b, consisting of category 2 conceptions for CQ2, includes the function of the kidneys together with digestion for discharging waste. In the preteaching phase, four students stated that waste is produced only "with the help of our kidneys," whereas, after teaching, nine students focused on the kidneys and digestive organs for the production of waste. This is because the experimental teacher linked the processes of digestion and excretion for waste removal rather than teaching excretion in isolation. Table 4.1 section c (CQ3) represents the categories and frequencies of the pre and postconceptions of the organs responsible for waste removal. According to the pre (N 10) and postteaching conception (N 23), the "anus" was considered an organ responsible for waste removal. Again, most students consider the anus as an organ in the digestive path for waste removal. Although the urinary system as a whole has dropped from ten to two as a waste removal medium, the kidney as an organ for waste removal has increased from six to 18. Despite teaching, as indicated by the response to CQ2 (see Table 4.1 section b), the knowledge of how kidneys function in excretion is minimal. Although it was not mentioned during the preteaching phase for CQ3, 18 statements refer to the skin as an organ for excretory function. Most students now perceive that the intestines and digestive system are not necessarily the organs for removing waste, thereby resulting in a significant dip in the results.

The Replacement of Everyday Language with Scientific Labels. In the preteaching exploration exercise, the everyday labels students used for feces are: "toilet," "bathroom," "motion," "products," and "shit." For example, note the expression containing the word "toilet" – "when you drink, when you eat it is formed into toilet and urine." Such colloquial or "street" words may not be the type of everyday language that Brown and Ryoo (2008) are referring to when they advocate that everyday language must be developed before introducing the scientific language to enhance its understanding.

In line with the tradition of phenomenographic research, teaching did not focus on eradicating children's everyday or "street" language because it is well-accepted for common use to function in their daily life. However, science discourse (with proper language labels) in the science classroom is important for socializing young learners into the scientific academic community (Brown & Ryoo, 2008; Wenger, 1998). Therefore, the teacher's specially designed instruction enabled students to perform relational talk or distinguish the contexts in which everyday talk and science talk are appropriate. She helped students to conduct the science talk with science labels in science class, thus inducing relational conceptual change. Students were helped to attach language labels such as "metabolism, metabolic wastes, and nitrogenous wastes" to their understanding. In fact, five students refer to waste as "nitrogenous waste" and three refer to it as "metabolic waste" (see Table 4.1 section a).

The Difference in the Complexity of Students' Responses from Pre to Postteaching. Developed from students' pre and postconceptions based on CQ2 (see Table 4.1 section b) are three phenomenographic categories of how waste is produced: (1) the eating of food and digestion; (2) the functions of the kidneys and digestive organs; and (3) the cellular process. Suitable examples from students' expressions are used to discuss the quality of statements from simple (sensory – eating) to complex (cellular process) because phenomenography is a map of the hierarchy of knowledge to illustrate the relationships among descriptive categories, as well as to visually represent the increasing levels of complexity in knowledge expression and representation (compare Figure 4.3a and b).

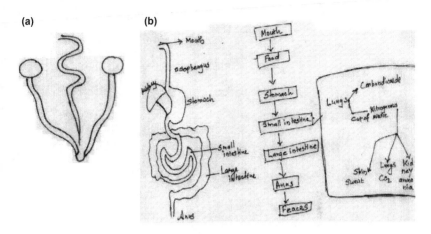

FIGURE 4.3 (a) A student's prior teaching representation of the urinary system (PEOE journal). (b) The same student's delayed post-teaching representation of the relationship between digestion and excretion (journal).

Phenomenographic Category 1: The eating of food and digestion. In the preteaching phase, some students focus on the initial step when they make simple statements such as: "By eating, the waste products will come out, and by eating they are produced." Within this category, students consider four conditions regarding how waste is produced. The first condition is: "They are produced *when we eat the food* the food goes to the stomach it is digested and it is pushed out." The second condition is: "They are produced *when we eat more food* that is not required for our body the leftover food becomes waste." The third condition is: "They are produced *when we eat the food the needed food is taken* and left is waste products." The fourth condition is: "They are produced *when we eat the food the good products are taken* and bad are excreted... first we eat the food it goes in the stomach and into the small intestine which takes the good products and the large intestine which take the bad products are excreted." This is evidence of how a student is confused with excretion and egestion. The student should have completed the statement as "the bad products are 'egested' or passed out from the anus."

In the postteaching phase, "Eating of food and digestion" ($N = 22$) received more emphasis. "The food is chewed in mouth and then it goes to stomach with the help of esophagus it goes into stomach and digestion take place food is absorbed in the small intestine and it separates solid and liquid the solid goes in large intestine and then it comes out as feces through anus" is an example of at least one student's conception. According to this student, solid and liquid separation does not take place in the small intestines. Such an explanation results in an obvious flaw in reasoning. While the minute flaw in children's reasoning is revealed in this article, in practice, we will not encourage sacrificing rich discourse for pointing out mistakes at the middle school level. A teacher, however, must be aware of students' flawed reasoning and bring these to their attention at an appropriate time. Students should be taught the scientific explanation that after the absorption of digested food by small intestines, the remaining undigested food enters the large intestine. At this stage, water is separated from the undigested food, thus making it solid, that is, feces. Solid feces is now stored in the rectum from where it is defecated.

Phenomenographic Category 2: The Functions of the Kidneys and Digestive Organs. In the preteaching phase, for CQ2, students talk about "the function of the kidneys along with digestion in discharging waste matter" (see Table 4.1 section b). The simplest version is: "By drinking water, urine is produced, eating food motion is produced" – the focus is on the physical processes of drinking and eating. The intermediate version is: "When the body absorbs the good materials in the small intestine the unwanted material is collected in urinary bladder and then excreted out by the anus. This waste is produced by kidney." In this version, the expression is partially coherent and somewhat flawed in terms of the theories and organs of egestion and excretion. The more sophisticated version is: "Mouth chewing take place stomach food is digested, small intestine, undigested food passes to large intestine throw out the waste products through anus as feces. Kidney removes urine." Teaching should aim for most students to acquire the more sophisticated understanding of waste production through a coherent, correct, and complete excretory model.

Notice the learning progression from phenomenographic category 1 (digestion) to phenomenographic category 2 (involvement of kidneys along with digestive organs) in the preteaching phase. Phenomenographic category 3, the cellular process, is the next level in the learning progression in the study of excretion.

Phenomenographic Category 3: The Cellular Process. In the postteaching phase, learning progression also includes the cellular process for waste production (see Table 4.1). One would not expect students to do science talk at this complexity, particularly at the middle school level. Three students, however, do talk about the cellular process. For example, a student explained that waste production includes the cellular process: "First we eat the food and it goes to the stomach by esophagus. It goes to stomach and goes to the small intestine where good products are absorbed then it goes to large intestine where water is absorbed. Then it goes to all parts of the body. It goes to the cell. There the nitrogenous waste is produced. Then it goes to the kidney and from there it goes to the urinary bladder and from there it goes to anus in the form of feces." Note the confusion in the last phrase revealing the connection between excretion and egestion. Relatively few students seem to have achieved the highest level (category 3) by including the cellular process to think about how waste is produced. Conceptual shift should be from category 1 to category 3 for CQ2, where the last category encompasses all aspects of waste production.

A number of students were able to recapitulate without any prior revision of the subject matter and draw diagrams in their delayed postteaching activity (after three months) similar to the PEOE activity, denoting the link between digestion, cell, and other excretory organs. A comparison of Figure 4.3a and b, the former representing the prior teaching drawing of one student and the latter representing the conceptualization by the same student after three months, suggests the value of conceptual change teaching.

Two years ago, I was drawing a honey bee using the five pencil method taught by artist Darrel Tank online. This method of drawing requires precision. I came to a deeper understanding of how engaging students not only to sketch ideas but producing close to naturalistic diagrams in biology or other relevant sciences. However, it takes much time and effort to develop realism in the knowledge construction phase. To me, it seems worthwhile. When I was drawing the honeybee, I learned so much about the anatomy of insects and light and shadow. This experience took me back to learning zoology as one of the two electives for three years at Lady Doak College. This course also had several hours for labs. The professors there insisted that we draw accurately labeled diagrams in our specialized journals. Examples were laboratory specimens, dissections of the frog's cranial nerves, jointed legs of the insects, and slides of fish scales viewed through a microscope. Each year of the B.Sc. program, the government examiners evaluated our laboratory journals. Drawing with precision might be another way of approaching science teaching and learning. Similarly, learning to play an instrument help to construct scientific knowledge. It is the teacher's ingenuity to bring all students to the banqueting table to manifest their talents.

4.7 CONSIDERATION FOR CONCEPTUAL CHANGE TEACHING ON EXCRETION BASED ON TEACHER ACTION

For students to understand the relationship between the digestive and excretory systems, it is important to systematically connect their existing knowledge about the digestive system to the excretory system rather than teaching these systems in isolation. Figure 4.3 clearly shows the connection that the teacher made between the digestive system and excretory system. What we learn from the decision and the action of the experimental teacher to teach both phenomena simultaneously and approach the study of excretion via the phenomenon of digestion is that the logical point of entry to a particular area of science and its interconnections should be kept in mind when planning to teach from a conceptual change perspective. In line with this indirect approach, the exploration task and questions should appropriately relate to learners' experience (simple and sensory), such as the ones used in this study (i.e., eating sandwich and drinking water). Then students' conceptions can be logically connected to a more complex phenomenon such as the cellular processes of excretion.

4.8 ARGUMENT FOR PHENOMENOGRAPHY AS A CONCEPTUAL CHANGE APPROACH

To design a lesson sequence that incorporates variations of students' conceptions of a science or natural phenomenon, phenomenography shows much intellectual empathy to the learner and learning. To demonstrate intellectual empathy, we did not categorize students or their ideas as high achievers, medium achievers, and low achievers and high, medium, and low, respectively. Rather, the teacher used phenomenographic categories as frameworks to develop the CKCM-based curriculum.

4.9 PEOE STRATEGY TO PROMOTE RELATIONAL CONCEPTUAL CHANGE

Phenomenographic categories of description can be subject to rigorous structural analysis and interpretation using other conceptual models, if that is the objective. For example, based on Chi and Roscoe's (2002) argument on conceptual change, students' conceptions in this study can be interpreted as "preconceptions" and not "misconceptions" that need to be repaired. If the teacher's exploratory question focused on excretory waste, and students had considered feces as excretory waste, then there would have been one misconception – incorrect categorization of feces as excretory waste. If so, ontological categorization is important for teaching purposes. Our notion is that students would have considered feces as excretory waste even if the question focused on the topic of excretion because students' emphasis was on feces for waste product and egestion as the process for which the anus serves as the organ for waste removal. This is not surprising because earlier studies on excretion corroborate this claim.

Based on this study, we advocate phenomenography for school science because it manifests intellectual empathy to the learners and to the variations of their ideas. The categories of ideas are not used as preconceptions or misconceptions, rather as variations. If it did, the teacher would have been consumed with altering or repairing misconceptions – a deficit approach to conceptual change. However, when phenomenographic categories are treated as fundamental curriculum frameworks and approached from the *Variation Theory of Learning*, relational conceptual change may occur.

4.10 RELEVANCE OF CKCM TO TRADITIONAL ASSESSMENTS

Amidst reform, achievement in science from a traditional perspective still occupies a central place in science education, and it will probably remain with us in the foreseeable future. As observed in this study, the reform-based CKCM has stood the test of traditional assessment. Therefore, it is important to use defensible models such as the CKCM that consist of exploration and categorization of learners' ideas and construction and negotiation of meanings that lead to better test results and conceptual understanding. For India's vision and mission of science education, it is imperative that schools adopt a relational conceptual change inquiry promoted by the CKCM. The relevance of CKCM transcends the Indian Education system, where "traditional" teaching methods still persist. Concerns related to "traditional" teaching methods are widespread globally. This being the case, it is imperative for teachers to consider using a teaching model such as the CKCM for science achievement and relational conceptual change, as pointed by this study.

The Flying Saucer

5

Dream to a Science Educator's Realities

5.1 THE FLYING SAUCER

I had a dream one night. There was a flying saucer parked in a big field. The commander, a middle-aged, tall, slender, handsome white man, wearing a gray suit studded with insignia and a hat approached me and asked me what the delay is. You are the main person. Your husband and son are aboard. You must be the first person to get in and take your seat. The captain led me carefully to the flying saucer. All around the capsule were rectangular sealed glass windows one foot apart. I took my place along with my husband and son. We were excited. Lifting gently, and gradually, the flying saucer launched into space. We were going up and up above the earth. I peered through the windows. I saw the dark earth lit with thousands of 60 Watt amber iridescent light bulbs turned upward to the sky. The view was beautiful. I asked the commander to fly further into space. He replied gently, we are above the earth, far into space, and that should be enough for now. This dream is still vivid. I was wondering what this dream meant and what the future holds. What is in store for me at Wayne State University (WSU)?

5.2 PAULA WOOD – "I WILL SERVE YOU WITH CRYSTAL WARE"

Before the dawn of 2000, I was looking for better opportunities. I was ready to leave the University of Manitoba. My well-wishers expressed, you are big fish in a small pool. I experienced racist attitudes from students and faculty. I was one of the three visible in color in the faculty. Winnipeg is a Christian belt and primarily a white society. There were several possibilities during the two years of my search for a suitable position. WSU had advertised for a science educator with a concentration in physical science for two successive years. I did not apply the first year. I waited for the second year to submit an application. I was invited for the interview. Maria Feriarra picked me up from the Detroit Metro Airport. Sharon, Norman, Gina, and Maria joined me for dinner that night. The only request I made to the group was that I will not be working on Sabbath. Immediately, Sharon responded that this is the kind of person we need here in Detroit, one who stands for conviction. My mind drifted to the interview at the University of Toronto, where they drilled me with religious concern. I was wondering why. They needed someone to teach evolution.

The next morning, my research presentation was on conceptual change from the history of science perspective. The interviews were with the science education committee, the assistant dean, the research dean, the dean, and human resources. I enjoyed talking to each of them. What the dean told me about attendance at meetings, selective salary, health insurance, and retirement funds are still vivid. I returned home that night. Very soon, the dean called me and offered me the position.

I told our son that I had a job offer in Detroit. He said, "Why there?" The people were amicable and helpful. The university was inviting, and I had a positive feeling. I mentioned this to some people at the Faculty of Education, University of Manitoba. One person told me, "You are going to your own kind." Another told me, "Their language research group is powerful." I was ready to move, and was asked to spend two days in Detroit.

Because my husband was overseas at the time, I was allowed to bring my son for a site visit. As soon as we came, he drove straight to the medical campus and was impressed at their facilities.

Millender Center in the heart of Detroit as a place to live pleased both of us. Facilities in The GM Tower and the River Walk are all aspects we liked. I lived there until 2012, although I looked around for single homes twice and returned to the same place. No housing could attract me. The Detroit River view and the Canadian view from my penthouse on the 33rd floor, the highest in Millender Center, was breathtaking. The fireworks from the Detroit River on Independence Day were dazzling.

Two nights before I left Winnipeg for Detroit, Dean Paula Wood called me again. She advised me that I don't have to pack anything. The movers will do everything for me. However, she warned me about crystalware and asked me to hand-pack myself with a lot of paper. It was then she told me I will serve you with crystalware. I was excited about her offer in all aspects. I was able to negotiate with her to bring a graduate student with two years of funding. My starting fund from the faculty and the central administration was lucrative. She told me that she boasted to the president that she was able to bring the best science educator. She also mentioned that there was one with more publications. Still, we wanted you because there was more balance to university roles and responsibilities. When I joined the faculty, and into my years of work, I would have liked the dean to communicate with me, but that never happened. I was, however, told that she and the faculty respected me highly. She did not forget to put me forward along with a few others to the Crain's business call. Neither was she quick to nominate me for the distinguished award before time (Figure 5.1).

FIGURE 5.1 Promotion to full professor with the WSU president (Irvin Reid, left), Jazlin (center), dean (Paula Wood, right).

FIGURE 5.2   Douglas Whitman (WSU College of Education dean) and Jazlin Ebenezer.

Nineteen years later, in 2019, the dean of College of Education, Professor Russel Douglas Whitman, nominated me for the Charles H. Gershensen Distinguished Faculty Fellow. I was the second College of Education recipient between 2001 and 2019 and the only one university-wide in 2019.

Jazlin Ebenezer professor of Science Education is selected as a Charles H. Gershenson Distinguished Faculty Fellow at Wayne State University (Figure 5.2).

5.3 VICE-PRESIDENT'S CONGRATULATORY COMMENTS

Congratulations to Dr. Jazlin Ebenezer, professor of Science Education in Teacher Education for the College of Education, who has been selected as a Charles H. Gershenson Distinguished Faculty Fellow at Wayne State University. Dr. Ebenezer award letter reads: "The Distinguished Faculty Fellowships have been established to recognize and provide support for members of the faculty whose achievements and current activities in scholarship, research, and/ or artistic performance and creativity continue to hold national distinction. Support for these fellowships was created by a special action of the [Wayne State University] Board of Governors designed to recognize and assist the intellectual pursuits of selected senior faculty members." Dr. Ebenezer will be recognized at the WSU's Academic Recognition Ceremony on Monday, April 25, 2019, at 4:00 p.m. at the McGregor Conference Center.

Vice-President's reading:

Professor Ebenezer is a nationally and internationally recognized thought leader and researcher in conceptual change inquiry in science education. Her influential research and innovation in innovative technology experiences for students and teachers often combine research with community engagement within Detroit. Professor Ebenezer is widely respected and invited as a keynote speaker for technology-embedded scientific inquiry throughout the United States and the world. In lieu of her exceptional scholarship, scholarly record of service, and influence of her international work, Wayne State University recognizes professor Ebenezer as a Charles H. Gershenson Distinguished Faculty Fellow (see Figure 5.3).

Many Facebook friends posted congratulatory notes:

Christianna Singh: Congratulations! You've done all of us at LDC very proud... just to think that part of your academic journey was with us...God bless! ☺

FIGURE 5.3 Jazlin with Keith Whitfield, Provost and Senior Vice-President for Academic Affairs.

Rajaratnam Abel: It's almost like a lifetime achievement award. Congratulations and God bless your efforts.

Robert B Bairagee: Congratulation Prof. We praise God for your outstanding academic contribution. May our loving God continually bless you to exalt His glory.

5.4 RESEARCH, TEACHING, AND SERVICE ARE ONE

At the University of Manitoba, I had two-pronged research: chemistry education and preservice learning to teach science. At WSU, my research on teaching and learning is the center of my work; however, the foci have shifted. The NSF-ITEST grant redirected my research to integrating technology. At this point, I am focusing on preparing and professionally developing teachers in the use of innovative technology on my initial chemistry education and preservice teacher preparation. Whatever my research focus might be, I believe and act on the premise that teaching and service revolve around research. For example, my conceptual change inquiry research and Technology-embedded Scientific Inquiry (TESI) model development are the foundations of all science courses I teach. I do not use any textbooks to teach. I tell my graduate students that I am the textbook. I expect my graduate students to read profusely based on the topics. Service should not be merely serving in one's faculty and the university. Serving the international community brings honor to the university. Selective salary recognition should involve equivalence of serving not only but also others. The College of Education should encourage faculty service for tenure, promotion, and selective salary purposes according to the interests and talents of its members.

My service to the international community involves the lessons I have learned from my classroom research. When researching is in my blood, it is not hard to talk about it anywhere and anytime, as was the precursor to the World Bank appointment to evaluate a seven-year $56M project. I believe teachers in preparation and those in school need to wear a research hat. They should study their teaching and their learning to make sensible decisions and take reasonable actions. For example, one of the science teachers showed me her record of marks of students of over two years and showed me how Common Knowledge Construction Model (CKCM) positively impacts science achievement. In Lynda Wood's study, the teacher tested the CKCM with alternative students and saw a significant difference. May Bluestein tested Computer-Assisted Instruction (CAI) in language fluency using Doll's curriculum design. There was a considerable difference in results. My only concern is that teacher–researchers should sustain a research mode. The argument for quality assurance calls for evidence-based impact of teaching on students.

5.5 LAKE ERIE WATERSHED PROBE WITH INNOVATIVE TECHNOLOGIES

When I first came to WSU, I was encouraged to work with the College of Science to secure extramural grants. Because I was successful in securing the Social Science Humanities Research Council of Canada grants twice, I wanted to write my proposal to the NSF. While contemplating this, the research dean asked me to work on a project that school districts had proposed to him. So, I served as the principal investigator to promote the idea they suggested. I offered the frameworks. The school districts and their supporters provided the practical aspects. The group from the school district was interested in submitting to the NSF-ITEST program. Once we got the grant, it was the school personnel who trained the teachers together with an engineer who trained teachers using a Geographic Information System (GIS) software. This section is elaborated in the next chapter, which is on integrating technology into science education.

5.6 WHO IS YOUR AUDIENCE? PRACTICES WHAT SHE PREACHES

I first met Ling Liang at an AERA meeting. She had recently graduated from Indiana State University and was attempting to understand academic life. She asked me about writing and publishing articles. One night, Ling called me and asked me whether I would be willing to go to China. She discussed the background of this call. I was ready to go to Nanjing University, Nanjing, China, on the invitation of Prof. J. Hao, and Prof. Liu. I prepared for the seminars like none other. I even sent the topics and the content that we will discuss. I conducted six seminars (30 hours) on teaching, learning, and research in science education to elementary and secondary teachers, as well as to science teacher educators.

My husband accompanied me to China. Our meeting was in a big hall for three days. Ling translated my academic talks from English to Chinese. Two graduate students helped me to converse with Professor Hao. We were treated to the best vegetarian food at every meal in a private room. I waited to eat rice with everything that was served. I was told that rice is for the poor, not for this group. But I had not eaten any Chinese food without rice or noodles as the base. Even a dish of noodles was not an item on the table.

After my initial introduction, I conducted the workshops. I introduced the relational conceptual change inquiry through hands-on activities to the 150 attendees. Neither notes nor PowerPoints were given. I asked the guests to make sense of what I was doing. At the end of the session, I gathered myself with the hope that everyone found my discussion about teaching and learning to be valuable.

Professor Hao called her two graduate students and wanted to speak with me. I was curious to know what she had to say. Through her graduate students, Professor Hao told me whether I know who is in front of me and who I am speaking to. I said I do not know. She noted that these individuals are from across China. They are sent by the government spending a lot of money. They are key school personnel who learn and then pass the information to local school science teachers.

That afternoon session on day one, I carried on with what I had planned. I did not change my approach to educating Chinese K-12 science educators. There were no PowerPoint presentations or notes scheduled. Again, as soon as the workshop was over, Professor Hao called the two graduate students and conveyed the same message, reiterating that I should give notes. Yet, I continued with my workshops with no change. Despite Professor Hao's dissatisfaction with how I approached the seminars, she treated us royally, providing us the best food and taking us on tours in the evenings. The hospitality was generous and remarkable. One of the graduate students often mentioned to me that Professor Hao is well-known and compelling in China. She can do anything she wants. I could not judge the graduate student's motive for giving me Professor Hao's background information. I took it to mean that Professor Hao will call me again because of her professional status.

The evening before the final meeting arrived. Professor Hao called her two graduate students by her side. She asked the students to tell me, "Tomorrow's attendance by the science educators will tell me whether the workshops were useful." I said to myself, typically, not too many attendees remain at meetings on the last day. So, if people leave, it is not my problem. And I did what I thought is the best for Chinese science educators. I used the approach of educating them that reflected my belief and practice.

The next day, I conducted my last workshop. The auditorium was full. I was so pleased because this was supposed to be a sign of my seminar's success. In the end, much time was given for the science educators to testify openly about what they learned during the seminars. For some reason, I was not concerned. I am used to face-to-face criticism, and I do not retaliate.

Many stood up and gave their testimonies. One of the graduate students translated each science educator's statement. All witnesses made positive remarks. I remember one account vividly. One science educator said, "I have attended so many workshops over the years as a leader in science education. This was the best because this expert practiced what she preached. Others merely talk, but she modeled." After the meeting, the science educators, more than usual, took pictures with me. There was also a professional photographer who took photos of the whole group. What a celebration. I am a celebrity!

The day was nearing for us to leave for India after the seminars in China. That evening, Professor Hao called the graduate students and asked them to take my husband and me to a very comfortable, well-equipped guest room. I was wondering what she was going to tell me next. Among the group was Professor Liu. The girls asked me whether I will be willing to go to Macau to teach summer school. I answered quickly without much thinking, "Oh, no, no, we are on our way to India for another engagement." Professor Hao, through her graduate students, immediately responded, "It is not immediately. It will be later, and the invitation will come through the department of education in Macau. They wanted me to conduct the summer school, and I have recommended you to the Macau officer." I was surprised. After all, finally, she understood what I was doing in the workshops. I told her that I will come. Next summer was our travel to Macau.

5.7 NO DROPOUTS WITHIN TWO WEEKS

As discussed with Professor Hao, the educator in charge of the summer school program, sent me an email message in 2004 and requested me to come. I observed that this educator was an alumnus of the University of Manitoba. So I was more enthusiastic. I sent all the information he required and got ready to go to Macau with my husband for three weeks. Professor Hao and the two graduate students came for a few days to Macau to help me settle.

From Hong Kong, we traveled by boat to Macau. Professor Hao introduced me to the 25 experienced science teachers who were taking summer school for 42 hours. The topic was *Integrated conceptual change science for secondary teachers: Teaching, learning, and assessment strategies.* I followed my SCE 5060 course syllabus for the secondary science methods course I teach at WSU. The science teachers did some work, but they did not seem interested. It appeared that they were forced to take the course. These science teachers would just walk out of class to answer their cellphones. Somedays, some teachers will not show up to class. The teachers would say that they had to be in school because the principal had called them for meetings. I was not used to such behavior. In the SCE 5060 methods course, the fundamental class to prepare science teachers would never behave like this. Once an engineer transfer student would walk out of my classroom in the United States as he wished. I told him he cannot be in my class if he continues with this behavior. I also reprimanded another student for continually texting in class. I told her that I will inform the field placement office about her conduct in the course and not place her in a school. A physics teacher from an affluent school was not fully engaged in an advanced graduate class. I happened to see him on my way to class that day. He told me that I was the first one to discuss his indifference in class with him. This young teacher had received NSF-PAEMST award that year for being one of the best secondary science teachers in the nation and the only one in Michigan.

Returning to my experience in Macau, I was frustrated. I felt Macau officials brought me to their schools, and I may not be helping the science teachers at all. While such thoughts were going through my mind, one day, the education director called me to his office to talk. I spent much time in his office discussing my feeling of not being able to reach out to the science teachers. After listening to me, he said that I was doing excellent work because I kept the course going and the teachers kept coming. Frequently, if teachers do not like summer school, they all drop out within two weeks, and there is no more class. This is our experience with summer schools. This is the first time we have a summer session for science teachers. My eyes were opened. He continued to state that Macau is a gambling city. It is Las Vegas of the East. The students do not care about academics. Their goal is to work in the casinos, where they are paid much more than a teacher. I left his office fully satisfied. For me, it was a cultural shock because Chinese students in the United States and Canada are very hardworking. They are achievers! Lo and behold, on the last day, every science teacher was there. They were excited to take pictures with me and wished me well. They also said they have learned how to teach science. Professor Hao returned to Macau to see me off. This time, she asked me whether she can translate my textbook in Chinese for science teachers. I promised her to write a new book in the context of China. A Chinese publishing company sent me contract papers. Since they were in Chinese I did not understand them. At the same time, I was going through the full professor rank. I wanted to write a textbook, but it never happened. Looking back, I feel sorry for missing a golden opportunity. Professor Hao is retired. I think about her often and of her influence in science education. At the point of writing this book, I am assessing a science education professor at the University of Macau. There is still an opportunity to meet Professor Hao again.

5.8 TAKE QUALITY ASSURANCE INTO YOUR HANDS

On a trip to Turkey in 2015, Professor Saka and I spent many hours to write an article on physics problem-solving. This article was based on the data he had collected in his physics methods class over several years. I also helped his wife write a paper on drama for biology education. So, I got to know both the husband and the wife. They took me to their village, and we traveled to the mountains, where I saw beautiful meadows of wildflowers.

After I returned to the United States, Professor Saka proposed to the scientific committee to invite me as the keynote speaker to the 6th Annual International Congress on Research in Education (ICRE): Quality Assurance in Education: Policies & Approaches at Recep Tayyip Erdogan University in Rize, Turkey, October 13–15, 2016. Referring to their governing documents and European materials on quality assurance, I talked about the arguments and practices for teacher preparation and education. I also brought to their attention the US models of accreditation, specifically the CAEP standards and procedures.

Prof. Cemil Öztürk from Marmara University asked me whether I would be willing to give the same talk at his University. As promised, professor Öztürk invited me as the keynote speaker to discuss the same subject at the International Teacher Training and Accreditation Congress, Yıldız Technical University, Istanbul, Turkey, in 2017.

My message in both conferences was not to wait for the government to initiate quality assurance. Instead, I advocated taking quality assurance into their hands and be committed to it. It should not come from an external force. Instead, it should be an everyday affair of internal bodies at the university. They know the standards and procedures the government expect. It is up to them to meet those standards. Quality assurance should be their ongoing work, not just in time collection of evidence. They should drive quality assurance, not the government.

5.9 EMPOWERING E-GENERATION VISION 2020

In 2014, Dr. Yashpal Netragaonka, assistant professor at the MAAER's MIT School of Education & Research, Pune, India wrote to me asking for help with the Scholarly Research Journal for Interdisciplinary Studies. I was willing to assist him in his scholarly pursuits. He invited me to be the foreign editor for the journal. After sending my curriculum vitae, I lost connection with him. In January 2016, I was at Spicer Adventist University, Pune. Professor Sheila Chacko invited me to be a keynote speaker at the EduSearch Conference. I was requested to speak on the topics of qualitative research and integrating technology into education.

When I was in Pune, I thought of Dr. Y.N., who had contacted me in 2014. It was an opportunity to see him because he is from Pune. But I had forgotten his name. After I returned to the United States, surprisingly, I heard from Dr. Y.N. He reminded me that we have had a previous discussion about academic exchange. He told me that he had included my name as the keynote speaker for the International Conference in India held in March 2016. The theme of the conference was Empowering E-Generation Vision 2020. He requested me to attend this conference.

Remember, in 2004, I went to Pune, India, and met Professor Sheila Chacko. At the time, I gave a seminar on research in science education to professors of sciences and educators. During the same visit, on invitation of Professor. Nanibala Immanuel of the same college, I gave an hour-long seminar on science education to graduate students, as well as a two-hour talk on science education to all education students. Subsequently, we worked on a research project.

Dr. Edison Samraj introduced me to this great visionary, Professor Vishwanath Karad, MIT founder and UNESCO chairholder in 2004. Dr. Karad lovingly calls Dr. Samraj his spiritual father. I had the opportunity to have a dialogue with the president and five scientists. Our conversation that day revolved around science and spirituality. Listening to Dr. Karad was spiritually uplifting. He took me on a tour of the World Peace Center, MAEER's MIT. He asked me to be a Special Guest of Honor and give a keynote address for the Third Session of the International Conference on *Ultimate Reality and Meaning of Life* to 300 scientists at *the MIT Dineshwar Hall.*

This is the second time I was at MIT invited by Professor Yashpal Netragaonkar. I gave the keynote address at the two-day *School of Education & Research International Conference: Empowering E-Generation Vision 2020* at the Dineshwar Hall on March 11, 2016. Although this invitation came independently as a result of my international scholarship in science education, I was not new to MIT because of my first introduction by Dr. Edison Samraj in 2004.

On the day of my talk, I was accompanied by Professor Vaibhav Jadhav, Pune University, to MIT. I was greeted in each office by an officer. Finally, Dr. Karad's daughter took me to her father's office. I was delighted to see Dr. Karad. His office is the size of a vast hall with a big desk and a kingly chair on one side and seats for guests on the other side. Around his office were pictures of all religions, including Christ. This scenario represented goodwill to all faiths.

When time came, I was escorted with significant others with pomp and glory to the same hall in which I had once addressed the scientists. The program started with the candle lighting ceremony, which I have never observed before. As the guest speaker, I had the privilege of lighting the oil lamp. The worship service, candlelight ceremony, welcome ceremony, and all talks were genuinely uplifting. Dr. Karad's recorded UNESCO speech was played at the opening ceremony. The address was touching and kindled my mind, heart, and soul. I am sure such was the experience of everyone who attended the Education and Research Conference. Dr. Yashpal Netragaonkar must be commended for organizing such a well-attended education and research conference. May God help him hold the flame of knowledge trim and burning. May he keep the spirit alive in the years to come (Figures 5.4 and 5.5).

I opened my remarks at MIT by stating that I have come back to India as their own (Figure 5.6). My keynote address was entitled "Engaging Students in Technology-Embedded Environmental Research Projects: Science, Story, and Spirit." The message was based on my $1.2M grant from the NSF, Washington, DC. That was a research project for the professional development of science, technology, engineering, and mathematics high school teachers over three years (2004–2007). My message addressed three modes of thought/space – science, story, and spirit – advocated by William Doll, a curriculum theorist to replace the modern linear curriculum with a nonlinear post-modern curriculum. A discussion on Doll's frameworks will be picked up in the final chapter. In line with the three modes of thought, I strongly urged the conference participants to integrate information communication technology in STEM education in India at the school level and in higher learning in teacher preparation and professional development.

FIGURE 5.4 Jazlin with Professor Vishwanath Karad, MIT founder and 2004 UNESCO chair holder.

FIGURE 5.5   Lighting ceremony at MIT.

FIGURE 5.6   Keynote address at MIT.

That afternoon, after lunch, I was hoping to attend sessions and be with the conference participants. Instead, Dr. Karad arranged for a car and an MIT officer to take my husband and me to visit their 150-acre campus in the suburb of Pune. I was awed by the facility, which housed a school, BEd college, and music college. Their medical college was soon to move to this campus. Dr. Karad's handiwork in the honor and memory of Raj Kapoor, a famous Indian Bollywood star, was elevating.

The principal of the school who we visited was cordial. I commend the men and women who work there, keeping the place immaculately clean. For example, I saw women cutting grass with a simple tool, and the lawns looked like a green carpet. The beauty and the spirit of the workers on the MIT campus are virtues to emulate.

Each building reflected a unique character and meaning. Dr. Karad's commitment, compassion, care, and concern toward all people must be heralded to the world. As we were driving back to Spicer Adventist University, I told my husband, "Dr. Karad's work is blessed because he puts God first." For he rises before the sun to spend one-hour in thoughtful prayer and meditation. This might be the secret of MIT's success. What a great example of a spiritual leader, leading an institute of that magnitude.

Since 1983, the MIT Group of Institutions has grown in leaps and bounds. MIT is housed in more than ten campuses in the state of Maharashtra. The MIT Group provides education in the fields of engineering, medicine, pharmacy, marine engineering, insurance, distance education, telecom management, lighting, design, food & technology, retail management, business administration, school of government, and school education. More than 50,000 students are pursuing various courses taught by more than 4,000 qualified and dedicated teachers. Under the banner of the World Peace Centre, MAEER's MIT, Pune, India, with an UNESCO Chair for Human Rights, Democracy, Peace, and Tolerance on 12th May 1998, has championed the cause of promoting value-based universal education system for spreading the message of peace in the society based on the appropriate blending of science, technology, and spirituality. MIT is world-renowned. Students from around the world aspire to study in this institution.

5.10 PhD DEGREE CONFIRMATION AFTER DISSEMINATION

Another surprise invitation was to Thailand. I forwarded the email invitation to our son, who, in turn, sent it to someone to make sure that this invitation is real and not spam. I was notified that this is a genuine call from Thailand.

Dear Jazlin,

My name is Jirarat, Assistant Dean for Academic and International Affairs. I email you about our visiting research scholarship at King Mongkut's Institute of Technology, Ladkrabang, Bangkok, which is in the top ten universities of Thailand. This grant allows a research collaboration for 3 months, in the field of educational technology, and covers the return flight ticket (economy class), accommodation (5,000 baht/month), and an allowance (50,000 baht/month).

Qualifications: candidates should have published at least 3 papers in the last 3 years and, in total, should have published more than 10 papers in renowned journals listed in the ISI database (as a correspondent author).

If you are interested in visiting us at King Mongkut's Institute for 3 months sometime this year, please don't hesitate to let me know, preferably as soon as possible.

Best regards,
Jirarat Sitthiworachart, PhD
Faculty of Industrial Education
King Mongkut's Institute of Technology Ladkrabang
1 Soi Chalongkrung 1, Ladkrabang, Bangkok
THAILAND 10520
Tel: +66 2329 8000 Ext. 6060
Fax: +66 2320 4511

Dear Jazlin,

This invitation is for the period of 3 months. During this period, you will engage in the following activities:
- lecture class(es) related to your research interest
- researcher work with King Mongkut's Instituter of Technology Ladkrabange (KMITL) researcher and lead to at least one ISI paper publication
- researcher mentor
- manuscript editor/reviewer
- modulator and commentator of graduate seminar
- invited speaker or trainer in a specific workshop or public seminar

By the way, we would like to invite you to be a keynote speaker at our international conference on Developing Real-Life Learning Experiences: Smart Education for Sustainable Development (the previous year link http://161.246.14.28/DRLE2016/) which will be held on 16th June of this year.

If you agree, please forward your bibliography, to place on our website, and some brief information on the topic about which you would present.

Thank you very much for your time, and hope to hear from you soon.

My work with Ji, her colleagues, and the graduate students were highly productive. Very soon, I learned that a PhD is conferred only when the candidates meet a condition. That the candidates publish either two articles in high-impact journals or publish one paper and present another paper at a scholarly conference. Thai students find these requirements challenging because of the language barrier.

At the time I went, there was a leadership PhD candidate who was attempting to publish an article in an international journal. She held an administrative post in the school system. She could not graduate until she had published an article in a high-impact scholarly journal. She was desperately seeking help because a journal had refused to accept her paper. She came with a translator to meet with me and begged me to help her get through her PhD. I promised to help her. However, writing the paper was not easy because of the language barrier. Over three months, besides doing other work, as listed above, I was continuously working with this PhD candidate.

We also kept in touch with her committee members to make sure they approve of what she is doing with me because they were co-authors. Unfortunately, the journal we sent the article to did not accept it. However, the journal editor wanted me to be on their editorial board. So, I took the opportunity to convey to them the conditions in which Thai PhD candidates work. Finally, I invited the dean's wife to help me communicate with the student and what she should be writing in her paper. We sent the article to the original journal she submitted. They sent the paper for review. The student thought the paper was accepted and conveyed a wrong message to the committee members. I intervened in the misinformation. I mentioned to the dean that the paper has not been accepted. I also wrote to the editor that we need to help her, and we are willing to make necessary revisions. To our satisfaction, the reviewers had recommended the paper. I requested the editor to send her a note stating that the article was accepted and will be published in a specific volume and issue of the journal. Soon the letter was submitted to the university. It was just in time of her deadline to complete her PhD. It was also the convocation. What a relief! She is a Catholic and heavily relied on prayer to get over this hurdle.

The dean said my name should be attached to the paper or I would not get credit from the university. As I had mentioned earlier, one requirement was that I help a researcher publish at least one article. The outcome of our persistent work was the following article:

Booranakit, N., Tungkunanan, P., Suntrayuth D., & Ebenezer, J. (2018). Good governance of Thai local educational management, *Asia-Pacific Social Science Review, 18*(1), 62–77.

I also helped other PhD students to write. I also helped Ji with her papers on hybrid technology and one on phenomenography of architecture students' conceptions of technology use. Both papers were not in-depth enough. To publish articles in scholarly journals, particularly on technology, it is imperative that we submit articles on how we use technology with students and the effect technology has on students' learning, instead of mere conceptions. In my opinion, what the faculty of education needs is to establish a research group and a person who can articulate the content in English and edit the work. The faculty members should also be active researchers and be able to guide students toward in-depth research for potential publication. The development of an educational model based on document analysis alone is not sufficient for PhD study.

5.11 FULBRIGHT SPECIALISTS YOU REACH

I received my Fulbright Specialist status for 2013–2018. My preference was to go to India. On our return from Pune to the United States, we had to take our flight from Mumbai. So we decided to stop in Mumbai to visit Professor Solomon Renati, professor of psychology at the college. We also met his wife at the college; they were working together on research projects. Our early morning flight from Mumbai to the United States allowed us to spend time with the family. They had met my husband on many occasions before I came to know them. My husband was keen on me meeting this couple and their daughter. During our conversation, I talked to them about my Fulbright Specialist scholarship, and I told them 2018 was the last year I could go anywhere. They were happy to host me.

Upon my return to the United States, I sent my host the contact information in India that was given to me by the Fulbright Commission to initiate the project. My host institution contacted me and requested that I serve on the project. Dr. Girish Kaul from The Fulbright Commission in the host country contacted me and asked that I serve on the project. Further, IIE-CIES/World Learning contacted me and requested that I serve on the project. With help from Fulbright Learning Group both in the United States (Lisa League and Carolyn Stokes) and India (Dr. Girish), we filled the form. My project's title was *Sustainable Development of Youth from STEM Perspective*. The dates of the project were from July 15 to August 26. Professor Renati got the necessary signatures from his college authorities and submitted the form. I needed a special research visa to go to India because I was going to work in Mumbai for 42 days. Getting a work visa was difficult. Because I am Indian by birth, the Indian visa officer asked me for proof of Indian citizenship and renunciation of Indian nationality. It was hard to convince the officer that I was never an Indian citizen because my parents immigrated to Sri Lanka when I was a baby. I was a Ceylonese holding a Ceylon passport. I had to give up the Ceylon citizenship and surrender my passport when I immigrated to Canada in 1972. Fulbright specialist application and Indian visa application processes were arduous. However, everything worked out on time to go to Mumbai.

Renati's family came to pick me up. Within a few days of my arrival in India, I had to get police clearance before moving from Mumbai and starting work or going to another place. I spent the first week at the YMCA facilities.

I was permitted to go to south India to give lectures in various colleges before the start of my work in Mumbai on June 16. While I was still in south India, professor Renati had an unexpected transfer. He moved from KBP College, Vashi, Navi Mumbai to Veer Wajekar Arts, Science and Commerce College almost at the end of June. I was told sudden assignments are typical.

There was no proper housing or hotels in Uran, where the college was situated. The first two days I traveled by bus with my host and his wife to the new college on rough roads, which took 1½ hour. I wanted to experience bus travel. I opted to stay in the host's two-bedroom apartment suite at Kharghar that Dr. Renati graciously offered me free of charge. I felt safe and secure. In fact, the host college is supposed to fund my accommodation, subsistence, and local travel. I picked up the travel cost by taxi each day for INR 1600 + 100 rupees for each additional hour + the value for extra miles for the journey from Khargar to Veer Wajekar Arts, Science and Commerce College Mahalan Vibhag, Navi Mumbai.

I enjoyed going to open markets where vendors sold fresh produce. I also liked the variety of fruits such as guava, mango, and bananas. I chipped in sometimes. Dr. Solomon's wife, Sree Latha, prepared delicious meals regularly. She made my life pleasant and persistently provided me with nutritious food. I particularly liked the way she cooked greens with a lot of freshly grated coconut. I also preferred the Kerala brown rice. Every Saturday, after church, as customary, they took me to a government-based exclusive restaurant, where they had Kerala cuisine. It was this family's custom to treat guests in this unique restaurant. In this facility, there is a store that sold Kerala herbal oil. In fact, Sree Latha is the one who introduced me to herbal oils. In another section of the same building, there was a Kerala specialty store, where they sold ornaments and other products from Kerala.

Sree Latha was an excellent support to her family as Professor Renati conveyed to me about his wife. Likewise, she demonstrated care and love to me. Sharon, their daughter, filled the registration forms for the police clearance and also helped me in other ways.

I gave several lectures on various topics in education. One was a seminar to professors on relational conceptual change inquiry-based CKCM of teaching and learning that I had developed and published. The emphasis was on exploring and categorizing students' ideas with everyday tasks and integrating their conceptions into the Mumbai University-stipulated curriculum as frameworks. I also gave seminars to professors and students on knowledge-generating and validating tools such as the Prediction–Observation–Explanation strategy, student modeling and connecting to the explanatory models, and types of knowledge in science. I also gave a workshop on concept mapping and Vee diagramming to several groups of students. Seminars on the integration of innovative technologies into science education was a topic of interest. In the psychology class, I taught students to write a research report with all the sections and subsections. I advised students on writing research reports and papers for publication. I also encouraged them to pursue advanced studies abroad. I promised to look for opportunities for chemistry students. I evaluated several third-year students' computer sciences projects. Students used software languages they had learned to develop websites with various interfaces for a higher institution, bank, school, business, etc.

I gave motivational talks on the following topic: leadership and entrepreneurship by taking ownership. I had bought nearly all of Dr. Abdul Kalam's books in his store when I visited his home in Rameswaram, Tamil Nadu. His vision for governance of India – freedom from corruption and call for India to become a developed nation by 2020. I based my lecture on Kalam's vision for the younger generation.

Besides giving lectures, I immersed in college activities. I read the Indian reform documents, college bulletin, government-prescribed syllabi, and textbooks for various courses. I also read information technology students' project reports. I enjoyed reading these reports on the development of apps because I was in the process of writing the script to develop the Students' Ideas App. I conversed with both professors and students. I observed many classrooms and laboratories. I listened to lectures in chemistry, computer science, mathematics, psychology, and zoology, as well as practical work in laboratories in the subdisciplines of chemistry for BSc years I, II, and III. I took an active part in student learning. A professor told me that they have to take the subject-matter government exam to teach at the university level.

The professors do realize that they are doing much chalk and talk, although they make it meaningful and exciting with diagrams, graphs, and equations. Students take copious notes. Laboratory learning is highly scripted and procedural. If students get accurate lab results, they are asked to repeat the tests. There was no discussion. This sort of teaching is not surprising because India is exam-oriented, and Mumbai University is no different. Professors need to cover the prescribed curriculum. The medium of instruction is primarily Marathi. Science, Math, Psychology, and IT classes are instructed in English. And immediately, the same information is repeated in Marathi. I also attended some of their staff meetings and special functions, although in Marathi.

The professors and staff were amiable and warm and liked my presence in their classrooms. The institution was supportive of my work. I was immersed in day-to-day activities with professors and students. Sometimes, chemistry professors shared their food with me. A student brought me sweetened lentil-filled flaky bread from home.

I took part in the two-day *International Conference on the Development of Wholistic Health and Wellbeing: Unifying Body, Mind and Spirit* on August 22–23, 2018, at Veer Wajeckar College of Science, Arts, and Commerce, Uran, Navi Mumbai, Maharashtra, India. Dr. Renati was the director of the conference. I gave a 40-minute valedictorian address at the *International Conference: Pythagorean Theorem as the Curriculum Foundation: An Architecture for the Holistic Development and Wellbeing of Youth.* This paper is the last chapter of my book. The ideas were well received by the conference participants who had come from various parts of India. Referring to my talk, Dr. Renati said she is used to giving speeches, and it was great. Others wanted to make sure that my talk was in the proceedings.

One day because of the labor strike in Maharashtra, there were no students at the college. So I took the opportunity to visit Uran. Uran is a coastal town at the tip of Navi Mumbai, Maharashtra. It is primarily a fishing and agriculture village. Migratory birds fly into Uran, and I saw their water habitat. I wanted to see how life is among the fishermen and farmers. On August 15, India Independence day, I visited the Gateway of India and Elephanta Caves. One day, I went around Bollywood. The driver took some pictures in front of their gates. In the chemistry department where I was most of the time, I was introduced to a BSc student who is naturally a comedian. I asked him what he was doing in a chemistry laboratory. I encouraged him to go for an audition. The student went, and he said they liked him and was called for another test. After I went to Bollywood homes and took some pictures outside their gates, I told him that I will have the privilege of visiting him in his luxury home.

The most important outcome of my project was the emphasis on returning the construction of knowledge to students, and students taking charge of their learning. In this line of thinking, I believe professors and students learnt several knowledge construction tools. The seminars reflected intellectual empathy, care, and critique. I was able to reach the mind, heart, and soul of the professors, students, and the service staff, including the security officers who greeted me each day. It is crucial to keep in mind to organically follow their agenda, interests, and needs in their own time and space, not yours. My keynote address rounded up the vital lessons I shared with the college faculty and students.

Not everything was rosy. I encountered some challenges that prevented me from successfully completing all project activities or achieving all outcomes. I wanted to do an in-depth treatment of the various models of teaching and learning. I wanted to give a seminar on argumentation. A chemistry professor wanted me to give a talk on women's improvement. The college requested me to help students with scientific research. However, time, circumstances, and culture of education did not permit me to do other activities because of the following challenges:

The covering of the test-driven curriculum was the main objective in teaching. Priority was not given to small peer group discussion and teacher–students' interpretive class discussion.

The college's focus was on the preparation for the international conference that was to happen before I left. This left little time to do anything else that I planned. However, I observed that collaboration among professors and staff was exemplary. The principal and community provided support to bring the conference activities to fruition. The brochure, proceedings, and all other materials were printed and distributed promptly. The conference was organic in two or three separate sessions.

The students not showing up for individual help with their research projects was another challenge. I understand that they are usually given a list of research project topics, and a small group of students or individuals select one of the issues from the list. This list was not prepared and distributed to students as expected. The students were lost when I asked them to identify their problems.

I planned to collaborate with my host institution in the future. I thought the college might be an excellent context to do the following activities.

The principal of the college asked me to write a report of my professional experience at the college for their annual college journal. I planned to follow-up about the study of teaching and learning with interested professors and students. I was also interested in submitting my conference talk to a chemical education journal for possible publication. I wanted to test an information technology-integrated unit on chemical kinetics using an app at the college. I intended to establish a link, both formal and informal, between WSU and my host institution in India. This is because some students were very eager to improve their English language skills. They also wanted to learn how to do research. Thus, my long-term goal was to help students in conducting research and writing reports. Editing students' work enables them to articulate their thoughts through scholarly writing. Writing an MOU was also a possibility. These were my objectives. Collaboration may have brought these goals into fruition.

The lesson I learned from my experience at Veer W. College was to go with the flow rather than following through one's agenda. Because of this reason, I was able to reach the administrators, professors, students, and service staff and to touch not only their mind but also heart and soul. When I first met the principal in his office, I thought that this was the place for me, even though the long travel to Uran was an unpleasant experience because of bumpy roads. However, it gave me a glimpse of what life is like for people who travel on those roads daily. Once the potholes are smoothened, highways are paved, and bridges are built, the drive to Uran would become picturesque with the Ghats.

World Learning was particular about me contacting them when I arrived back in the United States and my home. World Learning asked me to contact them if there is anything else I needed before departure from my host country. While in India, the American Embassy would inform me about inclement weather, strike, and other unusual activities.

5.12 FULBRIGHT SPECIALIST PRE-TOUR IN TAMIL NADU

Before leaving for India as a Fulbright Specialist to Veer Wajekar Arts, Science and Commerce College, Navi Mumbai, I asked permission from the Fulbright Commission to leave the United States early so that I can give lectures on science education at Lady Doak College in Madurai and nearby women's colleges.

I made Lady Doak College, Madurai, Tamil Nadu, India, my home between June 24 and July 15, 2018. I was invited by Dr. Christianna Singh, the principal of Lady Doak College. The principal was cordial. She and some professors gathered in her house to have supper on the first night. She accompanied us with her guitar to sing songs of praise. She also made sure I am comfortably settled in the guestroom. She arranged for a visit to the Meenakshi Temple and other historical and cultural places in Madurai.

She also arranged for a tour to Ooty, a hill station in Tamil Nadu. I was accompanied by a woman when I visited these places. We took a train ride in the hills and asked the driver to meet us at the last station at Ooty. The place I liked in Ooty was the thread garden. There was a huge room converted into a garden. Everything in the garden, the plants with flowers and leaves were woven with colorful embroidery thread. The unique idea was that no needle was used. I had the opportunity to meet the owner of the thread garden and take photos with him.

I had ten seminars on educational issues: Leadership; a conceptual change inquiry model of teaching and learning; Chemistry learning – Science, Story, and Spirit; Interdisciplinary research, student cabinet – leadership, administrative council. I delivered two inaugural addresses on biotechnology and chemistry on invitation by the heads of the respective departments. Besides, I had the opportunity to speak with a group of interdisciplinary researchers on women's learning. This group is funded and founded by Dr. Jazlin Ebenezer Research Fund, a Wider Church Ministries Endowment Fund, which was established in honor of the education I received at Lady Doak College in India. The Fund provides grants to small teams of multidisciplinary faculty members at Lady Doak College to conduct research on how women learn science, as well as to disseminate findings at national and international conferences, as determined by Lady Doak College. The market value on September 30, 2019, was $15,453.36. I enjoyed every minute of my stay at Lady Doak College. The faculty, staff, students, and workers were warm and friendly. The college is deemed Grade A according to NAAC Accreditation.

Before leaving to the United States for Fulbright work, Nabisa Thasneem, a doctoral student writing her dissertation on students with disabilities, arranged that I see the college where she studied. Thus, I was invited by Dr. S. Sumayaa, Principal of Thassim Beevi Abdul Kader College for Women [Autonomous & Reaccredited], Kilakarai, Tamil Nadu, India, between July 8 and 12, 2018. I provided three seminars – Fulbright programs, a conceptual change inquiry model of teaching and learning, and educational research. I was impressed with Dr. Sumayaa because of her government-funded research of natural, organic brown sugar crystals obtained from the local palmyrah palm manufactured by her patented equipment. I had a tour of this invention. She also bought me local sweets prepared with palmyrah sugar. The Kalu Dodol was delicious!

Dr. Sumayaa arranged for me to visit Rameswaram, a port city. In my days in Ceylon, I used to travel by ship with my parents and siblings from Talaimannar to Danuschkodi. Unfortunately, Danuschkodi went underwater; the whole village perished. Then, the ship went to Rameswaram. I like the local food available in the port of each country. So I wanted to visit Rameswaram. This place is the hometown of Dr. Abdul Kalam, former President of the Republic of India. Dr. Sumayaa arranged for me to meet Dr. Kalam's brother (102 years old) and his daughter living in the same house. I spent a lot of time talking with Dr. Kalam's niece, who resembles him. Upon her request, I got blessings from her father. She also gave me a book authored by Dr. Kalam. Adjacent to the building was a museum and a store with his artifacts and books. I bought all the books he had authored. On the way back to Madurai, we stopped at the memorial. It was impressive. I had the pleasure of stopping at Pamban Palam Bridge to see water. I used to travel by train from Rameswaram to Madura when I studied at Lady Doak College. I also happened to see this train on its tracks and drive along for a long distance. It appeared to be the same blue passenger train.

My final lecture was a one-day seminar on the conceptual change inquiry model of teaching, learning, and research at the Crescent College of Education for Women, Madurai, Tamil Nadu, India, on July 13, 2018. Dr. S. Mahdoom Ariffa, the principal, invited me. Being a teacher's college, the professors wanted to work with me. They asked me for videotapes of the model that I am advocating. Right now, the Students' Ideas App is in development. I hope this app will be useful for teachers globally.

HyperCard to Artificial Intelligence for Relational Learning

6

6.1 HYPERCARD ACTIVITIES: STUDENTS' CONCEPTIONS OF SOLUBILITY

I developed a HyperCard activity on the hydration of salt based on students' conceptions of solubility. High school students explained and represented salt in water as melting, the formation of a hydrate, chemical combination, and something to do with charges. They found the transition from the macroscopic (visible) to the submicroscopic (invisible) representation challenging. Awareness of students' conceptions of solubility was the reason I took classes on HyperCard at the University of British Columbia as soon as I completed the dissertation (Figure 6.1).

In 1992, I received University of Manitoba Research Grant Fund to inquire into teacher conceptualizations of chemical knowledge construction, a collaborative teacher–researcher project. With this fund, I used the solubility HyperCard that I had developed to investigate grade 11 chemistry students' learning of solution chemistry. These students appreciated the HyperCard activities designed in black and white in the most primitive form. Students were excited, and their expression was "Cool." I taught the grade 11 class for four days, exploring students' conceptions and using a HyperCard activity that showed stirring the

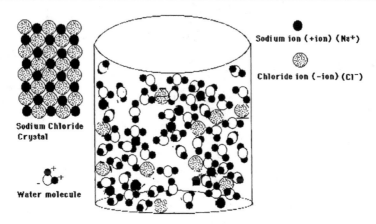

FIGURE 6.1   Animation – hydration of sodium and chloride ions.

salt in water. Students then used the HyperCard activity that demonstrates the hydration of salt, which is the hydration of salt. At the end of the fourth day, the teacher stated that he will complete the rest of the unit because he wanted to cover the syllabus. I was not happy. But there was no choice. Fortunately, I had collected data via students' work on the four days of teaching. I published an article in the *Journal of Science Education and Technology*. After a year, when I was photocopying in the room, a woman stepped in. She excitedly asked me, "Are you Dr. Ebenezer? I said, "Yes." Immediately, she replied that her daughter told her that she liked the way I taught chemistry.

Based on the chemistry classroom research with the HyperCard activity, I presented several papers at various conferences. In 1993, I presented a paper on *The use of Hypercard in developing common knowledge: An example in chemistry* at the international conference on Misconceptions in Mathematics and Science, Ithaca, New York. There was a large group in a big auditorium. An attendee said that I can make money out of my HyperCard activity. Another attached a price of $100. I did not expect this kind of response.

Although I took part in the conference and knew Dr. Joseph Novak personally, I did not believe that students' conceptions are misconceptions. I did not adopt the deficit path. The students' conceptions research strand in science education attached labels as misconceptions, alternative conceptions, and primitive ideas, to name a few. I viewed students' conceptions from the nature of science perspective. Why I would not refer to students' conceptions as misconceptions was asked in my doctoral oral defense.

In 1994, I received a *Social Sciences and Humanities Research Council of Canada* (SSHRCC) three-year research grant for *Science lesson sequences incorporating students' conceptions: A collaborative inquiry* to investigate chemistry learning. I collaborated with a chemistry teacher in her classroom. Based on this classroom research, I presented papers in seven conferences, namely, 1994 Canadian Society for Chemistry Conference and Exhibition in Winnipeg; 1995 International Conference on Distance Education, Moscow, Russia; CSSE in 1995 and 1996; International Organization of Science and Technology Education (IOSTE) Symposium in 1996, National Association for Research in Science Teaching (NARST) in 1997.

While at the conference in Cornell University, there was call for the 1995 technology conference at the University of Moscow, Russia, Soviet Union. I applied, and the paper got accepted. I made sure that I flew on the Aeroflot, a Russian airline. A group of students was traveling with me, and as the plane landed, they sang a song in Russian and then applauded with a shout of joy and excitement. I did not understand the language. However, I felt that students were thankful the plane landed safely and were happy to be back on Russian soil. I shared their joy and was pleased to be in a different country. I realized quickly that the conference was for administrators who were using technology in various ways in their universities and colleges. At the conference, a heart surgeon showed the audience a live heart surgery performed in his hospital in the United States.

I went to various sessions and appreciated the presentations. I was intrigued. No one talked to me, and I did not reach out either. I presented my paper entitled *A constructivist research-based digital solution chemistry unit: A proposed initiative for distance education* to a reasonably large group. After listening to the technical presentations, I thought there may not be much interest on the topic of my presentation. At the end of my presentation, many, both men and women, talked with me. Some hugged me. The attendees said it was one of the meaningful paper presentations on teaching and learning at the conference. They wished me well. The topic of my paper was appropriate, but I as an assistant professor, was with the wrong crowd.

In 1996, I presented a paper, *HyperCard environment for teaching and learning solution chemistry: A constructivist perspective*, at the 8th *International Organization of Science and Technology Education (IOSTE) Symposium*, in Edmonton, Alberta. Each presenter displayed a poster on the bulletin board on the eve of the day before the conference opening ceremony. The attendees took their place. The organizer of the conference introduced the keynote speaker and she took her place at the podium. She began her talk with my work and I was happily surprised. She used my work as an example to illustrate her points on using technology for teaching and learning. Referring to the title of the poster, the speaker stated that the design of the HyperCard was simple and focused. There were no bells and whistles. It was on teaching a particular concept in chemistry for understanding. She urged the audience not to make their technology design complicated. I left the conference with confidence that my HyperCard learning environment was excellent and not distractive with extraneous materials.

The research noted above culminated in a 2001 publication entitled *A hypermedia environment to explore and negotiate students' conceptions: Animation of the solution process of table salt* in *Journal of Science Education and Technology*.

6.2 CLASSROOM DISCOURSE ON WebCT – DISCUSSION BOARD

In line with my chemistry research I started as a doctoral student, I went back to classroom discourse. I thought the next work should be on classroom talk that develops students' understanding of solution chemistry. I contemplated on the possibilities of using the private platforms that science educators had developed for student dialogue. When the WebCourseTools management system was introduced at the University of Manitoba, I realized the Discussion Board will serve my research purpose because of the threaded and dated structures. The research data were readily accessible for analysis. While the WebCT was useful to the faculty in several ways, I appreciated the Discussion Board. I realized that there was no isolated communication tools that were available for conversations, dialogues, and argumentation. Examples of these were from OISSE, University of Toronto, and University of California, Berkeley, and I was interested in using these for my research on classroom discourse.

I first attempted to use the platform in my chemistry methods course. Secondary preservice teachers used the WebCT platform to learn how to teach chemistry. Based on the conversation data on the Discussion Board, I presented *Dialogues and arguments in project-based chemical inquiry: Research into the new dialectic* at the Biennial Chemistry Education Conference, The University of Michigan, Ann Arbor, Michigan. Along the same line of research and based on the research with another group of chemistry preservice teachers, I presented a paper on *Electronic discussion boards: Pre-service teachers' reflective dialogues on science teaching* at the *2002 IOSTE Conference*, Brazil, South America. This research led to my publication with three co-authors on *Community building through electronic discussion boards: Pre-service teachers' reflective dialogues on science teaching* in the *Journal of Science Education and Technology* in 2003.

Next, a middle school preservice teacher requested help with her practicum. Although I was not her advisor, I decided to help her in the classroom provided she does research in her practicum classroom. I did not realize that I was facing the same batch of students who Jennifer taught a unit of flotation in a Catholic school. I invited a chemistry preservice teacher to help me with research in the classroom. At the end of the practicum, I talked to the teacher about the students' understanding of the subject matter and insights on learning science. This exchange was based on the conversation I had with grade eight students at the end of the practicum. The focus of my research in the classroom was on grade eight dialogues on the WebCT Discussion Board, and not anything outside of it. I wished I had a recorder to capture all that the students said face-to-face in their last class period. As I contemplated on the missing opportunity of recording students' voices, what was most significant was what the grade eight teacher told me. First, she asked me whether I recognize the students. I said, "No." The teacher replied, "I have taught in this same school for 30 years, but I can say honestly that there has been no group of students as "smart" as these. "They are the same students you helped Jennifer teach when they were in grade two. This is why these students are exemplary." The teachers' statements were most rewarding. Based on this research with grade eight students and with some funds I received from Small 2002–2003 Research Grant. I presented a paper at *the Sixth International Conference on Computer Based Learning in Science* in Cyprus, July 5–10, 2003. The paper was entitled *WebCT dialogues on particle theory of matter: Presumptive reasoning in students' argumentation.* By the time I reached Turkey from Cyprus, the conference organizers invited me to publish the paper. The article appeared in a special issue titled *The Role of Research in Using Technology to Enhance Learning in Science* published in 2005. The Guest Editors were Zacharias C. Zacharia and Constantinos P. Constantinou.

In 2011, I assisted Ling Liang, a beginning professor to write a research paper based on the WebCT classroom discourse. The paper was entitled *Characteristics of Pre-Service Teachers' Online Discourse: The Study of Local Streams.* We submitted the article to the *Journal of Science Education and Technology.* The paper was accepted. Liang was not comfortable with the qualitative research. She did not want to pursue this research. Also, she did not want to revise the accepted paper and submit it to the journal. One day the editor wrote to her and asked Liang what happened to the paper. The editor encouraged her to submit it as soon as possible. Then Liang realized the significance of our paper. However, she did not want to pursue qualitative research. She told me this is the first and the last of this type of research. She found it painful to do something uncomfortable.

6.3 LAKE ERIE WATERSHEDS RESEARCH WITH INNOVATIVE TECHNOLOGIES

It was proverbial that I launch a research program at Wayne State University with funds from the National Science foundation (NSF) on Lake Erie Watersheds during 2004 and 2007. We had our first meeting with the

National Advisory Board. I started our conversation by showing the cover page of my hardcover book that had the photograph of a group of students near the water using probes and sensors. The committee members were all men. They did not show any outward enthusiasm while I was excited about it because the wish to work with technologies had come to fruition. The 1999 photograph pointed to the future event that will take place beginning 2004.

The NSF awarded nearly $1.2M for my Innovative Technology Experiences for Students and Teachers (ITEST) project known as Translating Information Technology into Classrooms (TITiC—pronounced as Tie-Tech): Teacher–Students' Research on Lake Erie Ecosystem. This project improved the knowledge of 45 high school science teachers and their students in Information Technology (IT). It supported science curricula by applying research and technical skills to the Lake Erie watersheds. The TITiC project engaged students in environmental research projects that used innovative technologies such as probes and sensors, GPS, and GIS. As a result of the NSF's multiyear grant, together with co-authors, I published three articles in premier journals. The 2011 article was on engaging students in environmental research projects: *Perceptions of fluency with innovative technologies and levels of scientific inquiry abilities* in the *Journal of Research in Science Teaching*. The 2012 co-authored article was on *One science teacher's professional development experience: A case study exploring changes in students' perceptions of their fluency with innovative technologies* in the *Journal of Science Education and Technology*.

6.4 THE TESI MODEL

During the NSF project, I was ready to develop a model that integrates technology into science education. I contemplated on the significant scientific inquiry abilities. I distilled the following investigational abilities: using scientific ideas to shape research, applying mathematical tools and statistical software in investigations, connecting evidence and explanation, and communicating claims and arguments. These abilities promote learning as conceptual, social, and technological. The abilities reflected the following three hallmarks of scientific inquiry: scientific conceptualization, scientific investigation, and scientific communication (see Figure 6.2).

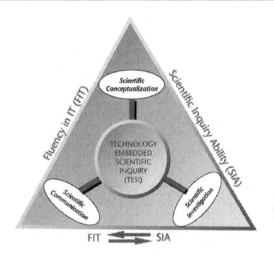

FIGURE 6.2   The technology-embedded scientific inquiry (TESI) model.

6.4.1 Scientific Conceptualization

Instructing students on scientific conceptualization calls for testing and clarifying conceptual ideas in ways that lead to a deeper understanding of subject matter. For example, *knowledge scaffolding and integration framework* developed by Marcia Linn in 2000 includes eliciting ideas, making ideas visible, adding ideas, developing criteria, and sorting out thoughts. Students are able to shape the problem of inquiry when they know the subject matter. Scientific conceptualization deepens when students engage in conducting empirical science with technology. Computer-simulated investigations, modeling, and visualization promote conceptual learning. Students have the opportunity to create chemical compounds using molecular modeling software to understand physical and chemical properties. They can also compare their representations.

Online concept maps promote collaborative learning. Students can discuss and negotiate the relational links among concepts of the knowledge network. Simulations such as molecular motion and the process of hydration help students learn concepts that are abstract or invisible. But students must reflect on the scientific model and engage in discussion. Learners themselves can animate the particle motion, for example, the three states of matter using software that use scripts. Scientific conceptualization via virtual models should accompany collaborative thinking and evaluation.

6.4.2 Scientific Investigation

Scaffolding students in scientific investigation involves students developing the following abilities: (a) formulating researchable questions or testable hypotheses; (b) demonstrating logical connections between scientific concepts guiding a theory and the design of an experiment; (c) designing and conducting scientific investigations; (d) using measurement instruments; (e) using mathematical tools and statistical software to collect, analyze, and display data in charts and graphs; (f) recognizing how investigation itself requires clarification of research questions, methods, comparisons, and explanations; and (g) weighing evidence using scientific criteria to defend illustrations and models (NRC, 1996). The Global Learning and Observations to Benefit the Environment (GLOBE) program provides an excellent example of ways in which above-mentioned abilities can be cultivated and fostered in a technology-embedded environment (see http://www.globe.gov/). In the GLOBE program, students immerse in authentic science investigations. They take scientifically valid measurements, analyze data, and report data through the Internet. Student publish their research projects based on GLOBE data and protocols. They create maps and graphs on a free interactive website. Students analyze datasets and collaborate with scientists and other GLOBE students around the world. GLOBE projects demonstrate the importance of scientific investigation. The projects also model the importance of the development of students' IT fluency and their scientific inquiry abilities.

6.4.3 Scientific Communication

Teaching scientific communication involves students in communicating research objectives, processes, results, and knowledge claims via classroom discourse and public presentation. These elements of research require critical response from both peers and experts. Communication technology tools have the capability to incorporate computer-based scaffolds to either support or refute competing theories. Position-taking is possible through constructing valid yet opposing arguments from multiple perspectives in response to issues. The emphasis on scientific communication represents a fundamental shift from teaching science as "exploration and experiment'" to teaching science as "argument and explanation" (NRC, 1996, p. 113).

The science education researchers value knowledge communication as much as knowledge creation. Students' communication with peers and teachers demonstrate that it is more than a passive activity in scientific endeavors. The online discourse formats enhance in and out-of-class communication and support collaborative group work. But the discourse on the critical examination of one another's viewpoints rooted in scientific inquiry should be vibrant. Learners require more opportunities to engage in scientific discourse. The discourse should accompany guidance in science classes.

6.5 TESI IN TURKEY

I have worked with Professor Muammer Çalik since 2006. He has translated my ideas into Turkish context and teaching preservice teachers. He and his co-authors published the article entitled *Improving science student teachers' self-perceptions of fluency with innovative technologies and scientific inquiry abilities* in the *Journal of Science Education and Technology*. Subsequently, this journal also carried another article by the same authors on the *Effects of 'environmental chemistry' elective course via technology embedded scientific inquiry model on some variables*. In 2018, as equal authors, we published a book chapter on *Innovative Technologies-Embedded Scientific Inquiry Practices: Socially Situated Cognition Theory*. Robert Zheng, Editor *Strategies for Deep Learning with Digital Technology: Theories and Practices in Education*. Nova Science Publishers, Inc.

TESI has a significant impact on the development of students' innovative technologies fluency and their scientific inquiry abilities. Little did I realize the significant relationships between the two until we completed the research. It was clear to me that the focus on the teaching of innovative technologies enables high school students to frame complex questions based on science-related social issues. Normally, we identify an issue, frame a question, and deploy the technology. Our research revealed the converse to be true as well (see Figure 6.2), that is, choosing a technology can identify socioscientific inquiry and pose a research question.

6.6 STUDENTS' IDEAS APP DESIGN AND DEVELOPMENT FOR RESEARCH

The Students' Ideas App (SIA) project began after I gave a keynote address on relational conceptual change inquiry at Lady Doak College. It is a Christian women's college in Madurai, Tamil Nadu, India. Dr. Pitchumani Angayarkanni

Sekaran, Associate Professor of Computer Science, expressed a desire to work with me in one of my projects. I was excited about the request because my research was on integrating innovative technologies into science education. Furthermore, US schools are emphasizing the importance of computer science in STEM education. I welcomed her request wholeheartedly. Since then, we have worked together laying the groundwork for the SIA project. However, it is challenging to work long distance on a complex project that rests on relational conceptual change inquiry.

To continue with the mission, the Charles H. Gershensen Distinguished Faculty Fellow funds of $24,000 I received in 2019 for two years are dedicated to the SIA project. The computer science professor came to Wayne State University on March 01, 2020 to design the SIA and pilot-test it in two Michigan science teachers' classrooms. She will be developing SIA with me until June 2020. The face-to-face conversations between the computer science professor and I were crucial because my experience in developing the SIA via Zoom or Skype was challenging. The difficulty was because the SIA design needs to depict the technology and assessment-embedded relational conceptual change inquiry approach to teaching and learning that integrates epistemic spaces for the "what," "where," "how," "why" of science. It requires the computer science professor to be with me physically to design and develop the SIA that aligns with the alternative orientation to learning. This interdisciplinary collaborative work demonstrates how and why it is significant that educators work with scholars in multidisciplinary fields. Such work helps to translate educational theories into practice, in this case, using machine learning to integrate students' ideas into the science curriculum.

The SIA research stems partly from the above multiyear NSF research project and accomplishments. A critical analysis of more than 100 students' scientific research papers in the TITiC project showed that of the 11 scientific inquiry National Standards (NRC, 1996), students were adept at six. They were weak in the five that pertained to communication in science epistemic spaces, namely, scientific conceptualization and scientific investigation. The study implied that the teachers' lack of knowledge in communication in science in the two overlapping areas manifested in students not knowing to incorporate the elements of communication in their scientific papers. Hence, it is essential to develop teacher–students' communication in science within the preceding two epistemic spaces. The conversation involves student writing, oral discussion, and visual representations using innovative technologies within a discourse community. Science teachers need to develop high school students' in-depth conceptual understanding through a relational conceptual change inquiry approach using scientific practices to become competent. Examples are framing questions, modeling, and arguing with evidence and explanation. Scaffolding theoretical knowledge and scientific practices are fundamental to engaging students in using the "what," "where," "why," and "how" of science in long-term, complex socioscientific issues.

Therefore, the approach to teaching in the SIA project involves a relational conceptual change inquiry approach. This approach consists of eliciting students' ideas by asking What do you think?, a second-order question, and explicitly incorporating those into the curriculum. The relational conceptual change inquiry approach also involves negotiating scientific explanations based on students' ideas through teacher-facilitated collaborative small group discussion and teacher–students' interpretive whole-class discussion. Moreover, it entails using hypermedia, computational modeling, and simulations. Finally, the approach translates the conceptual understating to shape socioscientific inquiries using the scientific practices outlined in the National Standards. In the investigation of socioscientific issues, apart from using the innovative technologies mentioned above, the SIA project aims to use virtual reality to develop student knowledge through superimposition of computer-generated images and information onto the real world. This approach shows promise in teaching and learning of science concepts. It also uses conceptual understanding to probe society-related problems framed by scientific practices. Once students develop theoretical knowledge and fluency in scientific practices, they can investigate socioscientific issues with innovative technologies mentioned above. This approach develops expertise and abilities, as well as dispositions to science and technology, which are the foundations for STEM higher learning and careers.

SIA is a vehicle for professional teacher learning. The students can also follow the teacher with SIA for their knowledge development. The participants will be able to install SIA on their iPhone or iPad or laptop. Teachers will use it as a model to design lesson sequences of their choice, as well as develop, implement, and evaluate it in their classrooms that characterize underrepresented youth. We will use real-world solution chemistry learning as one example in the SIA. This lesson sequence exemplifies different types of chemical knowledge (macroscopic, submicroscopic, and symbolic), increasing in complexity using the high school appropriate Internet modeling technologies. The lesson sequence also identifies and shapes a socioscientific problem of inquiry. It is related to the real-world solution chemistry unit that will deploy geospatial, data analytics, and communication tools. A relational conceptual change inquiry approach (Ebenezer, Chacko, Kaya, Koya, & Ebenezer, 2010; Wood, Ebenezer, & Boone, 2013) in conjunction with the TESI (Ebenezer, Kaya, & Ebenezer, 2011; Ebenezer, Columbus, Kaya, Zhang, & Ebenezer 2012, Ebenezer, Kaya, & Kassab, 2018) underpins the solution chemistry unit in the SIA. These evidence-based learning approaches underpinning the SIA have proven to develop in-depth scientific knowledge and dispositions needed to contribute to future STEM higher learning and workforce. Students will work with geospatial mentors and STEM experts while investigating the socioscientific issues, such as lead in water.

The computer science associate professor with the script and guidance from me designed the SIA-based e-content for experiential learning in solution chemistry from a relational conceptual change inquiry approach. For example, the students explore salt-water activity to generate their conceptions of dissolving – preassessment, as demonstrated in the Salt-Water Exploration Activity (see Slide 1 below). In this lesson, SIA will show to the teachers how to explore students' ideas of a natural phenomenon systematically using second-order questions (what do you think?). SIA banks students' ideas. SIA categorizes students' ideas and tabulates them. SIA directs students to engage in scientific inquiry activities with learning technologies available in SIA based on the categories of description (see Table 6.1 below for SIA's preceding functions).

The scientific conceptualization in SIA enables pre- and in-service teachers to explore their students' conceptual models of "dissolving" with a suitable everyday task. Through several activities with real or virtual laboratory investigations, 3D visuals, and animations, students will construct explanatory models. There will be an argument pattern so that students can map their argument and use it for interactive and argumentative discourse. Teachers and students, through interpretive discussions and negotiation, collaborate, co-create, and communicate to validate knowledge. This procedure is a way of achieving social objectivity. Negotiation connects the visible (macroscopic), invisible (submicroscopic), and symbolic (equations and math) knowledge of solution chemistry. Students find it challenging to navigate the types of knowledge when learning chemistry. There will be an embedded formative assessment for each activity. This learning space in SIA will also have reflections and reinforcements, as well as a summative assessment.

The socioscientific investigations, such as the ones conducted in my ITEST grant (2004–2007) described earlier, provide students the opportunities to do community-based inquiry projects both in and outside the classroom. The use of GIS and geospatial technologies enable the teachers and students to investigate problems in deep learning activities tied directly to the teacher and student community. The socioscientific issues converge interdisciplines, including social, cultural, and political. SIA points preservice and in-service teachers to the actual use of geospatial technologies. It also integrates decision-making and action-taking scenarios virtually that develop a better appreciation of the socioscientific issues.

Years of research suggest that the depiction of students' early conceptions are the same consistently. So, it is meaningful to develop SIA aligned with the Educators Evaluating the Quality of Instructional Products (EQUIP) rubric (Achieve, 2014) for professional teacher learning. In sum, the lesson sequence

TABLE 6.1   Categorization of students' conceptions

CATEGORIES OF DESCRIPTION	STUDENTS' EXPRESSIONS	SCIENTIFIC INQUIRY ACTIVITIES
Melting	Salt melted in water because I see a liquid	Students explore the difference between melting and dissolving with examples
Disappearing	Salt disappeared in water because I don't see salt anymore	Students do salt-water evaporation
Hydrating	Salt joined with water like this: $NaCl + H_2O \rightarrow NaCl \cdot H_2O$ This column has students' expressions—so words like "joined" has much meaning. So it is ok to leave as is.	Students heat copper sulfate crystals in a crucible and observe before and after heating. Students heat sodium chloride in a crucible. Students and the teacher discuss the difference between the two activities to show that sodium chloride is not a hydrate
Chemically Combining	Salt combined with water and produced a new substance $NaCl + H_2O \rightarrow NaOH + H_2$	Students explore several chemical combinations such as zinc in dilute hydrochloric acid to show that salt is not chemically combining
Charge Forming	Salt produced charges	Students use a conductivity apparatus in salt-water solution that makes the bulb glow to show charges are produced. Use a YouTube video to show the animation of the hydration process. Write an ionic equation for salt-water to produce sodium ion and chloride ion $NaCl_{(s)} + H_2O_{(l)} \rightarrow Na^{+1}_{(aq)} + Cl^{-1}_{(aq)}$

in SIA consists of exploration and categorization of students' conceptions, construction, and negotiation of scientists' explanations; translation and extension of students' understandings; and reflection and assessment. This teaching approach has the potential to transform science education so that pre and in-service teachers develop and adopt a useful model of teaching and learning in their classrooms. Students develop a strong foundation of knowledge and abilities for STEM higher learning and STEM careers. Because space is provided for students to document and interpret their responses about a specific science concept and related investigations and communication, SIA serves as one of the data collecting tools for research. The building of research structures in the SIA – the monitoring and tracking of students' conceptual understanding of scientific concepts and argumentative discourse, students' reflections of their conceptual understanding, development of scientific practices, embedded assessments, and grades – will enhance future research on learning and achievement.

The science of learning proposed by the Committee on Developments in the Science of Learning in 2000 is the basis for the SIA framework of science learning. Central to science learning is conceptual, epistemological, and functional. Underlying these are changes that have occurred in our understanding of the multiple aspects of science learning: (1) change in behavior explained by mental states, (2) mental functioning, (3) socially constructed knowledge, and (4) technological (i.e., embedding technologies that allow for learning flexibility). Science learning involves student exploration, construction, and translation of conceptions of natural phenomena; students' models of situations; and mathematical modeling, manipulation, and simulation. Vital is how teachers can facilitate the refinement of those conceptions and models by exposing students to additional data, and through discourse with conceptualizers and modelers. SIA enhances the teaching and learning of complex science from a relational conceptual change inquiry approach advanced by my research. The practical side of the forward position is the exploration of students' ideas of a natural or socioscientific phenomenon upon which rests the other aspects of learning.

Exploration of natural and or social phenomenon adopts the relational conceptual change inquiry learning theory advanced by Marton and Tsui (2004). This theory emphasizes the relationship between the conceptualizer and the phenomenon. It describes the possible conceptual variations individuals ascribe to a phenomenon. The practice of the relational conceptual change inquiry approach shows much intellectual empathy for the learner, which our students need.

SIA aims to develop teacher preparation and development. It maintains a balance between learning to teach science in the university classroom and teaching in the school classroom. The developed SIA ensures integration and coherence among the various components of a Teacher Education Program. The SIA is expected to equip the aspirant school teacher with the requisite scientific knowledge and understanding, abilities in scientific practices, and the scientific dispositions to address the challenges of becoming an effective science teacher. SIA is relevant to methods courses in all areas of social sciences and humanities in which all student teachers learn to teach.

Engineering for Learning Science Embedded in Societal Issues

7

7.1 STEM INTEGRATION

Since 2012, the New Generation Science Standards through reform efforts have been focusing on science learning in the context of societal issues that integrates science, technology, engineering, and mathematics (STEM). Thus, it is essential for science teacher educators to prepare or professionally develop science teachers in STEM education. We do know that most sciences are typically taught in isolation. The focus is on educating students in each subject area. Science learning is not situated in issues. Such researchers, as Angie Calabrese-Barton, push issue-based education, which integrates STEM. However, issue-based learning and integrating STEM seem to be reserved for an occasional program. It is not the norm in our schools. In research and practice of STEM integration in societal issues, the intention is not to lose the disciplinary focus. Too often, teachers focus more on the social aspects of issues and lose the science that initially framed the issue. Concurrent development of science content or the use of science content to study an issue takes second place. Often science gets neglected in the integration of STEM in the study of a social issue. I use a donut metaphor to help my students understand that studying a community-based subject may be similar to a donut not filled with crème. My students get the point. In the current school system, STEM and issue-based learning are rare. Thus, for at least one unit per semester teachers must engage students in the experience of creatively working through a societal problem and developing skills involved in STEM integration.

Researchers claim that the integration of STEM in issue-based teaching and learning reaches most students. The traditionally underrepresented groups – ethnic minorities, females, students with disabilities, and English language learners – are the beneficiaries.

Research reveals that engineering design experience in solving a societal problem proves useful in improving STEM learning in two ways. First, engineering design develops engineering habits and 21st-century skills. Second, student interest in STEM subjects grows. Engineering can help integrate STEM disciplines and provide a meaningful context to use scientific and mathematical concepts and principles that support STEM literacy. I believe that K-12 preparation for engineering instruction is crucial for the sustainable growth of engineering industry in the global economy. Thus, I reached out to Michelle Grimm, an engineering colleague at Wayne State University at the time, to help me with such thinking.

Grimm pointed out that bioengineering demonstrates the interdisciplinary nature of many engineering applications in today's society. She highlighted bioengineering attracts females – who represent only 10% of working engineers – to the study of engineering. An analysis of a bioengineering challenge shows that it can easily support the Next Generation Science Standards. Teaching from a bioengineering perspective develops students' knowledge, ability, and self-efficacy in (1) scientific and engineering practices, as well as (2) disciplinary core ideas related to engineering, technology, and life sciences. The National Research Council's focus on engineering and its decisive role in STEM education underscores a critical need for research-based science curricula that incorporate engineering in K-12 science. Such concentration promotes generative places of learning for minority and low-income students.

7.2 BIOENGINEERING AS AN EXEMPLAR

Bioengineering is one way of applying engineering to the solution of societal problems. We need to frame the bioengineering curriculum or any curriculum that involves STEM learning with the sciences of learning. Most notable are the conceptual, epistemological, and functional learning. Underlying these changes involve the understanding of multiple aspects of STEM learning. Mental states, mental functioning, socially constructed knowledge, and technology allow for learning to explain the shift in behavior or the way we act. STEM learning constituents are student exploration, construction, and translation of conceptions of natural phenomena, students' models of situations, as well as mathematical modeling, manipulation of materials, and simulation. It is vital how teachers can facilitate the refinement of students' conceptions and models by exposing them to additional data through relevant activities and discourse with conceptualizers and modelers.

The engineering design cycle, in some manner, involves defining the problem, identifying goals and constraints, generating multiple solutions, communicating ideas, and evaluating selected solutions. Incorporating the engineering design cycle to problem-based learning has been shown to foster creativity and improve the understanding of concepts among students. Valuing creativity, innovation, and problem-solving through engineering earlier in students' education can make students globally competitive. The engineering design cycle parallels the learning: exploration, construction, reflection, revision, and communication.

7.3 A POST-MODERN OUTLOOK FOR ENGINEERING

I have used Doll's post-modern curriculum frameworks developed in 1993. The four elements – richness, recursion, relation, and rigor – underpin the exploration, construction, and translation of concepts successfully in STEM settings with students, including the English learners and the disadvantaged.

When searching for textbooks on curriculum studies in my collection of books, I came across William Doll's book on post-modern curriculum. I had not opened this book since buying at an AERA conference. Among all textbooks on curriculum studies, I was drawn to Doll's position on curriculum theory.

William Doll comes from a biological perspective, which is organic and open, and not a closed system like Ralph Tyler's model of curriculum that the whole world uses. Learning, according to the latter, is preset. Doll's 2012 book has served as the centerpiece for our discussion in a curriculum study course for some years. Doctoral students appreciate Doll's thoughts on curriculum issues. Still, they also point out that the current state of affairs does not allow for such learning that promotes romance, generalization, and precision expressed by Whitehead. It is the romance stage in education that is missed in schools and exceptional education. I believe curriculum should be at the interface between the students and the subject. Bernard Ricca, a science educator, who also knows William Doll, explained to me this curriculum theorist's four Rs in one of our conversations.

Richness means the curriculum enabling students to actively engage in the discipline via exploration, construction, and translation. **Recursion** refers to the ability of the curriculum to operate, using a similar process, across multiple scales (e.g., from the individual through classroom level and to the disciplinary canon), and for the individual to be self-reflective. **Relation** explains the need to consider the ecology in which a classroom topic finds itself. The ecology of the classroom includes the learner, the teacher, and the disciplinary content under study, other disciplines, and other factors, such as the needs of English learners and students with disabilities. **Rigor** requires that the classroom approach parallels the approach of the subject in both process and product, although at smaller scales of importance, depth, and time. I have devoted Chapter 8 to detail Doll's curriculum frameworks.

7.4 BIOENGINEERING CHALLENGES

An environmental engineer, Dr. Allison Harris, provided me with examples such as wastewater treatment and biofuels that we can use with middle school learners. Michelle Grimm suggested to me engineering and medicine and engineering and physiology for high school students. These, she said, we use along with potential engineering challenge projects and associated biology and engineering content. I believe that the curriculum should include relationships to society, environment, and other disciplines, consistent with Doll's postmodern approach noted above.

Wastewater Treatment: Microbiology to Investigate Option is an example. For wastewater treatment, we may use microorganisms and determine the best growth and reproduction. Biofuels is another bioengineering challenge that involves biological energy conversion dynamics. Research points to conventional feedstocks, which may develop possible design solutions for biofuels. The study of the role of photosynthesis in energy flow is a strand. Interaction of biofuel crops with the environment may be a focus. Energy crops versus waste feedstocks energy flow also are of significance in the study of biofuels. In both examples noted above, students have an opportunity to define the problem, compare solutions, and observe how technology increases the number of solutions.

We may use bioengineering challenges. For example, Engineering and Medicine is a context to discuss how engineering supports medicine. Students can develop a timeline of medical technological innovation. Students define a problem that medical technology addresses and determine the constraints and goals of the system. Emphasis can be on researching and reflecting on how medical technologies influence challenges related to reducing healthcare costs. These bioengineering learning contexts offer students the opportunities to careers related to biology; scientific and engineering practice; compare the structure and function of a human body system or subsystem to a nonliving system; scientific reflection and social implications; products, processes, and policies; defining problems, define and research, technological issues, as well as making links among engineering, technology, science, and society.

Engineering and physiological systems compare the structure and function of a human body system or subsystem to a nonliving system, obtaining, evaluating, and communicating information. The engineering and physiological systems involve modeling systems, as well as identifying needs; establishing goals and constraints; developing solutions; and obtaining, evaluating, and communicating information.

An example illustrates how a bioengineering curriculum may support all learners. The above curriculum requires students to understand and design a technology to assist malfunctioning vascular systems. To develop the science concepts that underpin the design, students must first explore, construct, and translate their ideas. For this, students can watch a video of the functioning of a healthy vascular system and of a constricted one. Students should annotate and interpret that video on an interactive whiteboard or a low-cost storyboard based on their personal ideas. This annotation encompasses learning that benefits all students, especially students with disabilities.

Students discussing their initial models of vascular function based on the scientific content that they viewed on the video allows the construction of a more in-depth understanding. This learning, contrasting the initial models with explanatory models, aligns with the *richness* requirement of the postmodern curriculum. This step ultimately translates students' conceptualization to develop a problem statement that can be solved by an engineered system.

In this learning process, English learners can be paired with stronger English-speaking students. The National Science Teach Association in 2014 put it as follows: Seeking out and building on student-centered cultural and linguistic means fueled learning. Instead, our current practice focuses on academic language goals, often framed as vocabulary or discrete elements of grammar. While there are many other ways in which the various frameworks can be integrated, this example is indicative of the alignment between the national standards and structures of the bioengineering curriculum.

Pythagorean Theorem

8

An Architecture for Curriculum Design

8.1 WILLIAM DOLL

An assignment I give my graduate students is to translate post-modern curriculum frameworks to a discipline. They are expected to write a ten-page scholarly paper explaining Doll's post-modern curriculum frameworks (4R, 3S, and 5C) with persuasive examples taken from their discipline. This assignment provides students the opportunities to make sense of the post-modern curriculum in light of their practice, which usually follows a modernist view of the curriculum. Such is the education accorded to teachers at the graduate level. This assignment serves as a platform to disrupt the perspectives and practices of modernity (Figure 8.1).

8.2 PYTHAGOREAN THEOREM

William Doll advances a post-modern curriculum that is open and flexible as opposed to closed and stable reflecting modernism. A post-modern curriculum is in a state of flux, alternating between equilibrium $\rightarrow\leftarrow$ disequilibrium that leads to transformation (Doll, 2012). I visualize Doll's post-modern curricular frameworks to depict the Pythagorean Theorem, which states the sum of the

FIGURE 8.1   With Dr. William Doll and his wife, Donna Trueit.

squares on the sides of a right triangle is equal to the square on the hypotenuse (see Figure 8.2).

I use the Pythagorean Theorem as a metaphor to illustrate Doll's (2012) post-modern curriculum frameworks, which are 4R—Richness, Recursion, Relations, and Rigor; 3S—Science, Story, and Spirit; and 5C—Currere, Complexity, Cosmology, Conversation, and Community. Thus, the square of the 5C (Currere, Complexity, Cosmology, Conversation, and Community) equals the sum of the square of 4R (Richness, Relations, Recursion, and Rigor) and the square of 3S (Science, Story, and Spirit). Substituting $A^2 = B^2 + C^2$ in Figure 8.2 with the post-modern curriculum frameworks, we get $5C^2 = 4R^2 + 3S^2$ (see Figure 8.3).

Curricula founded on the 4R, 3S, and 5C are apt to develop the multiple intelligences of the learner – behavioral, intellectual, social, emotional, spiritual, technological, and political, thus contributing to the education of the whole

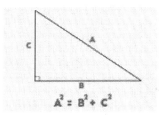

FIGURE 8.2 Pythagorean Theorem.

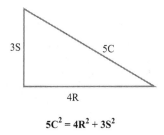

$$5C^2 = 4R^2 + 3S^2$$

FIGURE 8.3 Post-modern curriculum frameworks (Doll, 2012).

being – spirit, soul, and body. The soul comprises emotions, will, and intellect. Next, I explain each of Doll's frameworks with an exemplar in chemistry – the concept of dissolving, using the Common Knowledge Construction Model of teaching and learning.

8.3 THE FOUR Rs

Richness refers to the depth of a curriculum that holds sufficient problems, perturbations, and paradoxes (Doll, 2008). Learning experience emphasizes ambiguity and alternative perspectives. Knowledge contains layers of meanings, alternative interpretations, and infinitive possibilities that allow students to construct, co-construct, and reconstruct. The rules are simple and flexible encouraging diversity. Curriculum richness enables students to take complex and creative paths. A constrained curriculum is homogenous, predetermined, and uniform. When a unit of study consists of exploring students' conceptions of chemical phenomenon such as dissolving and incorporates the categories of description as frameworks into the curriculum, the study of the unit on solution chemistry characterizes richness. The concept of dissolving is problematic and creates perturbation and paradox in the minds of the learners. For example, students theorize dissolving as melting, disappearing, hydrating, chemically combining, and charge-producing when they observe salt added to a beaker of water and stirred. Such reasoning is prevalent because the salt-solution chemical system generates ambiguity and alternative thinking. Exploration of students' conceptions provides the opportunity to recursively consider meanings and interpretations from the explanatory ideals of chemists.

Recursion refers to the "looping back to what one has already seen and done" (Doll, 2008, p. 9). That is, the curriculum must have opportunities for students to revisit the initial concepts they learned and translate them to shape personal and societal inquiries. Such looping, thoughts on thoughts, defines human awareness. Looping is the process by which individuals make meaning when they reflectively interact with the environment and culture. Therefore, recursion is not a repeat of an event. The framework of recursion is reflective and open, developing competence and creativity. It provides a stairway to move to an upper level of learning. This space enables inventing novel ideas with information and innovative technologies. Students have already observed the salt-solution chemical system. Based on their conceptions of dissolving, students loop back, conduct inquiry with technologies in small groups, and seek evidence to explain the salt-solution system argumentatively. Based on the tests, evidence, and discourse (written and oral), students reflectively think about the chemical system from the experts' viewpoint.

Relations point to the student →← teacher; student →← community; novice ←→ expert; student ←→ student; and student →← simulation/text interactions to develop behavioral (biological), mental, social, emotional, spiritual, technological, and political intelligences. The relations with self, with text, with others, and with the local community foster development. Relationality is dynamic; elements are constantly active, making horizontal and vertical movements to create new possibilities. To study the salt-solution chemical system, after small group peer discussion, both students and the teacher can conduct an interpretive discussion after each test that generates evidence to explain dissolving. Students should also interact with simulated resources and textual materials on the dissociation and the hydration process of salt in water. The understanding of the solution process leads to curriculum rigor.

Rigor refers to a curriculum that is replete with thought-provoking opportunities, creating students who are consistently exploring and searching for new combinations, interpretations, and models (Varbelow, 2012, p. 93). This complex learning implies that a rigorous curriculum nurtures students' critical thinking through making connections, conceptualizing alternatives, and identifying possibilities. As Lewis (2004) articulates, learning embodies rigor when students understand how information or a particular skill in some manner translates to their lives. The students follow their thinking rather than merely achieving a preset learning goal. Understanding the solution process gives students many opportunities to explore the world of food, environment, and medicine – anything that is in the soluble or solution state.

8.4 THE THREE Ss

Science is the first mode of thought or space which is quantitative. It characterizes verifiable analysis, logical proof, sound argument, and empirical discovery guided by hypothesis. Good teaching, research or thinking = scientific, logical, rational, empirical, and factual. Only what is observed or observable and measurable is real. According to Bruner (1986), this mode of thought is "heartless" because it seeks to transcend the particular by higher and higher reach for abstraction. Alfred N. Whitehead points out that abstraction is "one-eyed;" it is tunnel vision. Merely learning the definitions of solution and the components of solution, as well as the symbolic abstractions of solution chemistry does not have much value.

Story is the second mode of thought or space, which is qualitative. It characterizes the story/stories of the particular; it is experiential and personal; and it is contextual, imaginative, and playful. Good teaching, research, or thinking = stories of the particular element(s), its history, culture, and environment. Each element/event/occasion is involved in a nexus of connections that are personal, historical, and cultural. No object is unto itself – each is contextual, and each is contingent. There is no such thing as the passive existence or single location. Whitehead states that reality is born through the interaction of each element or person. When exploring the world of solutions through standards-based scientific practices (NRC, 2012), students should keep a journal to chronicle their experience, discuss the contexts in which they are practicing, and narrate the lived-out experience and story of their knowledge generation, validation, and dissemination. It is only after the recursive contemplation of their understanding of knowledge production, the writing of the formal report occurs. In this nexus of complex learning, students are the creators of knowledge.

Spirit is the third mode of thought or space, which involves complexity. It characterizes action because of the renewal and regeneration of one's spirit. Good teaching, research, thinking = spiritual, creativity, and vitality. The spirit, the animator, the inspirer, and the creator gives life to the education process. It is the power that shapes the quality or mode of that situation or entity. The spirit is necessary to keep knowledge alive. Because of the findings in a real context, be it lab or in the environment, students should take the next step, which is taking actions. Perhaps, inform the community about their findings of the quality of water because of soluble substances and the effects it has on organisms, including humans. Students may also engage in reaching out to a larger sphere through publication in the chemistry department scholarly journal and actually submit to an academic journal. When students participate in complex activities, they become more spiritual.

8.5 THE FIVE Cs

This section presents Meldrum's (2008) profound explanations of the 5Cs.

Currere implies that curriculum is not about a teacher running a course. Preferably, the experience that results from students running the course (Doll, 2005). The curriculum is not about the subject, instead it is about each student's unique experience with the issue he/she relates with his/her life. However, learning does not depend on students' exciting experiences. Instead, learning happens only when a student reflects upon his/her exciting experience. Thus, the experience of the person does not count. What matters most is the secondary, the reflective act of knowledge that makes meaning and develops understanding, leading to the transformation of an individual. Hence, the teacher needs to work with the students and pose challenging questions, negotiate spaces, and help them interpret and theorize. This kind of education means that the experience is not what happens to someone, instead it is "what someone does with what happens" (Grumet, 1976, quoted in Doll, 2002, p. 44).

Complexity incorporates the world. The focus is on the domain, not its discipline. For example, the focus is on scientific practices that capture scientists' entire endeavor, not merely the scientific inquiry. Students need to face simple levels of complex interactions, for it is from these new and more complex levels arise (Doll, 2002). However, students are not left alone to chart new territories. Teachers and community provide a supportive learning environment for some stability. Tosey (2002) suggests that the "edge of chaos" (the dynamic between stability and instability or order and chaos) is the place to remain for learning to occur. If the learner faces too much order with no chaos, there is no creativity. If the student has too much chaos, it is like an engine misfiring. Creating order is necessary for pistons to be going up in the right direction, for the sparks to be happening at the right time for it to move. The edge of chaos is an essential space for becoming a well-rounded doctor, an engineer, a scientist, a teacher, a political leader, a diplomat, an entrepreneur, or anything one wishes.

Cosmology is typically the study of the universe and its history, focusing on "particles." Whitehead argues for replacing "particles" with a post-modern cosmology in which "actual entities" or "actual occasions" are "the final reals" of the world, and "these actual entities are drops of experience, complex, and interdependent" (Doll, 2002, p. 47). Cosmology is the integration of a harmonious balance of the science, with the story and the spirit. Florida (2005) argues that "human diversity and their contributions to the knowledge development is not merely enjoyable; it is essential for the health and well-being of an economic system" (p. 35).

Conversation for Doll (2002) is "one that captures us." When we actively participate or immerse ourselves in such a conversation, we need to speak clearly as well as listen carefully. "We need to hear back, recursively, both our own words and those of others" (p. 49). Pinar (2004) sees curriculum as an extraordinarily "complicated conversation" (p. 186). Trueit (2005) notes conversation as "the stream of thought and life" (p. 77) in open and fluid spaces, where there is chaos. Lachs and Lachs (2002) suggest that "within interactions, participants in the conversation normally reformulate their questions and their answers that they have not used before and thereby contribute to the creation of novel ideas and insights" (p. 233). Tlumak observes (2002) that the discussion approach to teaching generates "a critical, reflective, and dynamic learning environment" (p. 177). Doll (2002) adds that our hopes for convergence and transformation lie in conversation (p. 49).

Community means "What is most important to each of us is what we have in common with others" [...], that "they" really are "one of us'" (Doll, 2002, p. 50). Florida (2003) argues for building a creative community, "cities need a people climate even more than they need a business climate" (p. 283). The three Ts of creative places, according to Florida, are technology, talent, and tolerance (p. 249). Wenger (1998), believing that tolerance allows for diversity, argues that our imagination enables us to adopt or adapt to different perspectives regardless of space and time and accepting "otherness" conveys its "own language" (p. 217). Seitz (2003) tells us to "nurture self-expression within communities" (p. 385). Miller (2002) adds that dialogue is "the soul of education" (p. 97), and the "philosophy of interdependence complements dialogue" (p. 110). Doll (2002) argues that the concept of community encourages transformation where there is both "care and critique" (p. 50). Teachers help students to negotiate their constructs with themselves and their peers (Doll, 1993). Our educational system is competitive. Instead, it should be collaborative within a community – working with someone, sharing knowledge, and developing good ideas, excellent products, and functional design.

The questions every college teacher should ask are: What is knowledge? What is learning? When we answer these questions from the perspectives of knowledge generation, validation, and representation from various disciplines – the humanities, sciences, commerce, and arts –we will approach the same prescribed syllabi and textbooks in alternative ways, perhaps from the post-modern perspectives discussed above. It will be well for our heart and soul to practice the proper methods of teaching to enable our youth of the 21st century to develop the mental, social, emotional, and spiritual intelligences for their wellbeing. Thus, I urge every professor or lecturer to redesign and re-enact the study of the university cognitive-based curriculum with a focus on the 4R, 3S, and 5C—the Pythagorean Theorem.

Bibliography

Barton, J., & Collins, A. (1993). Portfolios in teacher education. *Journal of Teacher Education, 44*, 200–210.

Booranakit, N., Tungkunanan, P., Suntrayuth, D., & Ebenezer, J. V. (2018). Good governance of Thai local educational management. *Asia-Pacific Social Science Review, 18*(1), 62–77.

Bruner, J. (1986). Two modes of thought. In *Actual minds, possible worlds* (pp. 11–34). Cambridge, MA: Harvard University Press.

Çalik, M., Ayas, A., & Ebenezer, J. V. (2005). A review of solution chemistry studies: Insights into students' conceptions. *Journal of Science Education and Technology, 14*(1), 29–50. Top *h* cited research: 70 citations.

Çalık, M., Ayas, A., & Ebenezer, J. V. (2009). Analogical reasoning for understanding solution rates: Students' conceptual change and chemical explanations. *Research in Science & Technological Education, 27*(3), 283–308.

Çalık, M., & Ebenezer, J. V. (Equal authors, 2018). Innovative technologies-embedded scientific inquiry practices: Socially situated cognition theory. In R. Zheng (Ed.), *Strategies for deep learning with digital technology: Theories and practices in education*. New York: Nova Science Publishers, Inc.

Çalik, M., Ebenezer, J. V., Özsevgeç, T., Artun, H. & Küçük, Z. (2014). Improving science student teachers' self-perceptions of fluency with innovative technologies and scientific inquiry abilities. *Journal of Science Education and Technology, 24*(4), 448–460.

Çalik, M., Özsevgeç, T., Ebenezer, J. V., Artun, H., & Küçük, Z. (2014). Effects of 'environmental chemistry' elective course via technology embedded scientific inquiry model on some variables. *Journal of Science Education and Technology, 23*(3), 412–430.

Çalik, M., Ozsevgec, T., Kucuk, Z., Aytar, A., Artun, H., Kolayali, T., Kiryek, Z., Ultay, N., Turan, B., Ebenezer, J. V., & Costu, B. (2012). Analyzing senior science student teachers' environmental research projects of scientific inquiry: A preliminary study. *Procedia – Social and Behavioral Sciences, 46*, 379–383.

Chi, M. T. H., & Roscoe, R. D. (2002). The processes and challenges of conceptual change. In M. Limon & L. Mason (Eds.), *Reconsidering conceptual change: Issues in theory and practice* (pp. 3–27) Dordrecht: Kluwer Academic Publishers.

Collins, A. (1992). Portfolios for science education: Issues in purpose, structure, and authenticity. *Science Education, 76*, 451–463.

Doll, W. E. (1993). *A post-modern perspective on curriculum*. New York: Teachers College Press.

Doll, W. E. (2002). *Ghosts and the curriculum*. In W. E. Doll & N. Gough (Ed.), *Curriculum visions* (pp. 23–70). New York: Peter Lang.

Doll, W. E. (2005). The culture of method. In W. E. Doll, M. J. Fleener, D. Trueit, & J. St. Julien (Ed.), *Chaos, complexity, curriculum, and culture: A conversation* (pp. 21–75). New York: Peter Lang.

Doll, W. E. (2008). Looking back to the future: A recursive retrospective. *Journal of the Canadian Association for Curriculum Studies, 6*(1), 3–20.

Doll, W. E. (2012). *Pragmatism, post-modernism, and complexity theory: The "fascinating imaginative realm" of William E. Doll, Jr.* (D. Trueit, Ed.). New York: Routledge.

Duschl, R. A. (2003). Assessment of inquiry. In J. M. Atkin & J. E. Coffey (Eds.), *Everyday assessment in the science classroom* (pp. 41–59). Washington, DC: National Science Teachers Association Press.

Duschl, R. A., & Gitomer, D. H. (1991). Epistemological perspectives on conceptual change: Implications for educational practice. *Journal of Research in Science Teaching, 28*, 839–858.

Duschl, R. A., & Osborne, J. (2002). Supporting and promoting argumentation discourse in science education. *Studies in Science Education, 38*, 39–72.

Ebenezer, J. V. (1992). Making chemistry learning more meaningful. *Journal of Chemical Education, 69*, 464–467.

Ebenezer, J. V. (1996a). Christian preservice teachers' practical arguments in a science curriculum and instruction course. *Science Education, 80*(4), 437–456.

Ebenezer, J. V. (1996b). Developing common knowledge: A perspective of measuring success in science class. *The Manitoba Science Teacher* (SAG Special Edition), 38(2), 23–27. (Invited article by the editor of the Science Teacher Association of Manitoba). Reprint of this article in *Catalyst*, British Columbia, Canada.

Ebenezer, J. V. (2001). A hypermedia environment to explore and negotiate students' conceptions: Animation of the solution process of table salt. *Journal of Science Education and Technology, 10*(1), 73.

Ebenezer, J. V., & Achtemichuk, L. (1992). What do children really think of shadows?: A preservice teacher's explorations of science instruction. *The Manitoba Science Teacher, 33*(3), 11–15.

Ebenezer, J. V., Chacko, S., Kaya, O., Koya, K., & Ebenezer, D. L. (2010). The effects of common knowledge construction model sequence of lessons on science achievement and relational conceptual change. *Journal of Research in Science Teaching, 47*(1), 25–46.

Ebenezer, J. V., Columbus, R., Kaya, O. N., Zhang, L., & Ebenezer, D. L. (2012). One science teacher's professional development experience: A case study exploring changes in students' perceptions of their fluency with innovative technologies. *Journal of Science Education and Technology, 21*, 22–37.

Ebenezer, J. V., & Connor, S. (1998). Learning to teach science: A model for the 21 century. Upper Saddle River, NJ: Prentice-Hall, Inc., Simon and Schuster/A Viacom Company.

Ebenezer, J. V., & Connor, S. (1999, Canadian Edition). *Learning to teach science: A model for the 21 century.* Scarborough, ON: Prentice Hall Allyn & Bacon, A Division of Simon & Schuster/A Viacom Company.

Ebenezer, J. V., & Erickson, G. (1996). Chemistry students' conceptions of solubility. *Science Education, 80*(2), 181–201. (This article was requested by The Archives of Jean Piaget, University of Geneva).

Ebenezer, J. V., & Gaskell, P. J. (1995). Relational conceptual change in solution chemistry. *Science Education, 79*(1), 1–17.

Ebenezer, J. V., & Haggerty, S. (1999). Becoming secondary school science teachers: Preservice teachers as researchers. Upper Saddle River, NJ: Prentice-Hall, Inc., Simon and Schuster/A Viacom Company.

Ebenezer, J. V., & Hay, A. (1995). Preservice teachers' meaning making of science instruction: A case study in Manitoba. *International Journal of Science Education, 17*(1), 93–105.

Ebenezer, J. V., Kaya, O. N., & Ebenezer, D. L. (2011). Engaging students in environmental research projects: Perceptions of fluency with innovative technologies and levels of scientific inquiry abilities. *Journal of Research in Science Teaching, 48*(1), 94–116.

Ebenezer, J. V., Kaya, O. N., & Kassab, D. (2018) High school students' reasons for their science dispositions: Community-based innovative technologies-embedded environmental research projects. *Research in Science Education, 1*. doi:10.1007/s11165-018-9771-2.

Ebenezer, J. V., & Landry, A. (1994). Fostering common knowledge through problemsolving. *The Manitoba Science Teacher, 35*(2), 13–23.

Ebenezer, J. V., & Puvirajah, A. (2005). WebCT dialogues on particle theory of matter: Presumptive reasoning schemes. *Educational Research and Evaluation: An International Journal on Theory and Practice, 11*(6), 561–589 (Special Issue: The Role of Research in Using Technology to Enhance Learning in Science. Guest Editors: Zacharias C. Zacharia and Constantinos P. Constantinou).

Ebenezer, J. V., & Zoller, U. (1993a). Grade 10 students' perceptions of and attitudes toward science teaching and school science. *Journal of Research and Science Teaching, 30*(2), 175–186.

Ebenezer, J. V., & Zoller, U. (1993b). The no change in high school students' attitudes toward science in a period of change: A probe into the case of British Columbia. *School Science and Mathematics, 93*(2), 96–102.

Edwards, D., & Mercer, N. (1987). *Common knowledge: The development of understanding in the classroom.* London: Methuen.

Feher, E., & Rice, K. (1986). Shadow shapes. *Science and Children, 24,* 6–9.

Florida, R. (2005). *The flight of the creative class.* New York: Harper Business.

Ivarsson, J., Schoultz, J., & Saljo, R. (2002). Map reading versus mind reading: Revisiting children's understanding of the shape of the earth. In M. Limon & L. Mason (Eds.), *Reconsidering conceptual change: Issues in theory and practice* (pp. 77–99). Dordrecht: Kluwer Academic Publishers.

John 4: 4-15, Bible.

Lachs, J., & Lachs, S. M. (2002). Education in the twenty-first century. In J. Mills (Ed.), *A pedagogy of becoming* (pp. 219–228). Amsterdam: Rodopsi.

Lave, J., & Wenger, E. (1991). Situated learning: Legitimate peripheral participation. Cambridge: Cambridge University Press.

Lewis, N. S. (2004). The intersection of post-modernity and classroom practice. *Teacher Education Quarterly Summer,* pp. 119–134.

Liang, L., Chen, S., Chen, X., Kaya, O. N., Adams, A. D., Macklin, M., & Ebenezer, J. (2009). Preservice teachers' views about nature of scientific knowledge development: An international collaborative study. *International Journal of Science and Mathematics Education, 7,* 987–1012.

Liang, L. L., Sufen, C., Chen, X., Kaya, O. N., Adams, A. D., Macklin, M., & Ebenezer, J. (2008). Assessing pre-service elementary teachers' views on the nature of scientific knowledge: A dual-response instrument. *Asia-Pacific Forum on Science Learning and Teaching, 9*(1), 2–19.

Linder, C. (1993). A challenge to conceptual change. *Science Education, 77,* 293–300.

Linder, C., & Marshall, D. (2003). Reflection and phenomenography: Towards theoretical and educational development possibilities. *Learning and Instruction, 13,* 271–284.

Liu, X. (2004). Using concept mapping for assessing and promoting relational conceptual change. *Science Education, 88,* 373–396.

Liu, X., & Ebenezer, J. V. (2002). Descriptive categories and structural characteristics of students' conceptions: An exploration of relations. *Research in Science and Technological Education, 20*(1), 112–131.

Liu, X., Ebenezer, J. V., & Fraser, D. (2002). Structural characteristics of university engineering students' conceptions of energy. *Journal of Research in Science Teaching, 39,* 1–19.

Marton, F. (1981). Phenomenography—Describing conceptions of the world around us. *Instructional Science, 10,* 177–200.

Marton, F., & Booth, S. (1997). *Learning and awareness.* Mahwah, NJ: Lawrence Erlbaum Associates, Publishers.

Marton, F., & Tsui, A. (Eds.). (2004). *Classroom discourse and the space of learning.* Mahwah, NJ: Lawrence Erlbaum.

Meldrum, R. J. (2008). A curriculum for entrepreneurial creativity and resourcefulness in New Zealand. Unpublished Doctoral Dissertation, Deakin University.

Micari, M., Light, G., Calkins, S., & Streitwieser, B. (2007). Assessment beyond performance: Phenomenography in educational evaluation [Electronic version]. *American Journal of Evaluation, 28,* 458–476.

Miller, G. D. (2002). Abolishing educational welfare: Redrawing the lines of interdependency through dialogue. In J. Mills (Ed.), *A pedagogy of becoming* (pp. 93–113). Amsterdam: Rodopi.

National Research Council. (1996). *National science education standards.* Washington, DC: National Academy Press.

Novak, J. D. (2002). Meaningful learning: The essential factor for conceptual change in limited or inappropriate propositional hierarchies leading to empowerment of learners. *Science Education, 86,* 548–571.

Novak, J. D., & Gowin, B. (1984). *Learning how to learn.* Cambridge: Cambridge University Press, 199 pages.

Osborne, J., Simon, S., & Collins, S. (2003). Attitudes towards science: A review of the literature and its implications. *International Journal of Science Education, 25,* 1049–1079.

Piaget, J. (1973). *The child's conception of the world.* St. Albans: Paladin.

Piaget, J., & Inhelder, B. (1974). *The child's construction of quantities.* London: Routledge and Kegan Paul.

Pinar, W. F. (2004). *What is curriculum theory?* Mahwah, NJ: Lawrence Erlbaum.

Pinar, W. F. (2008). Curriculum theory since 1950: Crisis, reconceptualization, internationalization. In F. Connelly, M. He, & J. Phillion (Eds.), *The SAGE handbook of curriculum and instruction* (pp. 491–514). Thousand Oaks, CA: SAGE Publications, Inc.

Pinar, W. F. (2013). The reconceptualization of curriculum studies. In D. J. Flinders & S. J. Thornton (Eds.), *The curriculum studies readers* (pp. 149–156). New York: Routledge.

Pinar, W. F., & Grumet, M. R. (2014). A poor curriculum (3rd ed.). Kingston, NY: Educator's International Press, Incorporated, 271 pages (Grumet, 1976, quoted in Doll, 2002, p. 44).

Posner, G. J., Strike, K. A., Hewson, P. W., & Gertzog, W. A. (1982). Accommodation of a scientific conception: Towards a theory of conceptual change. *Science Education, 66,* 211–227.

Reilly, C., & Ebenezer, J. V. (1995). The POE strategy for chemical knowledge construction. *The Manitoba Science Teacher, 37*(2), 12–16.

Ruiz-Primo, M., & Furtak, E. (2007). Exploring teachers' informal formative assessment practices and students understanding in the context of scientific inquiry. *Journal of Research in Science Teaching, 44,* 57–84.

Saljo, R. (1988). Learning in educational settings: Methods of inquiry. In P. Ramsden (Ed.), *Improving learning: New perspectives.* London: Kogan Page.

Sampson, V., & Clark, D. (2008). Assessment of the ways students generate arguments in science education: Current perspectives and recommendations for future directions. *Science Education, 92,* 447–472.

Seitz, J. A. (2003). The political economy of creativity. *Creativity Research Journal, 15*(4), 385–392.

Thagard, P. (1992). *Conceptual revolutions.* Princeton: Princeton University Press.

Tlumak, J. (2002). Teaching through discussion. In J. Mills (Ed.), *A pedagogy of becoming* (pp. 177–198). Amsterdam: Rodopi.

Tosey, P. (2002). *Teaching at the edge of chaos.* York: Learning and Teaching Support Network Generic Centre.

Toulmin, S. E. (1972). *Human understanding.* Oxford: Clarendon Press, 520 pages.

Trueit, D. (2005). Water courses: From poetic to Poietic. In W. E. Doll, M. J. Fleener, D. Trueit, & J. St. Julien (Ed.), *Chaos, complexity, curriculum, and culture: A conversation* (pp. 77–99). New York: Peter Lang.

Tyler, R. W. (1949). *Basic principles of curriculum and instruction.* Chicago: The University of Chicago Press.

Varbelow, S. (2012). Instruction, curriculum, and society: Iterations based on the ideas of William Doll. *International Journal of Instruction, 5*(1), 87–89.

Wagner, T. (2008). Rigor redefined. *Educational Leadership, 66*(2), 20–24.

Walton, D. N. (1996). *Argumentation schemes for presumptive reasoning.* https://books.google.com›books

Wenger, E. (1998). *Communities of practice. Learning, meaning, and identity.* Cambridge: Cambridge University Press.

White, R. T., & Gunstone, R. F. (1992). *Probing understanding.* London: Falmer Press.

Wood, L., Ebenezer, J., & Boone, R. (2013). Effects of an intellectually caring model on urban African American alternative high school students' conceptual change and achievement. *Chemistry Education Research and Practice, 14*(4), 390–407. doi:10.1039/C3RP00021D.

Woods, P., & Hammersley, M. (1980). *School experience.* London: Croom Helm.

Zoller, U., Ebenezer, J. V., Morely, K., Paras, S., Sandberg, V., West, C., Wolthers, T., & Tan, S. H. (1990). Goal attainment in science-technology-society (S/T/S) education and reality: The case of British Columbia. *Science Education, 74*(1), 19–36.

Index